Mastering Machine Learning with Python in Six Steps

A Practical Implementation Guide to Predictive Data Analytics Using Python

Second Edition

Manohar Swamynathan

Apress®

Mastering Machine Learning with Python in Six Steps

Manohar Swamynathan
Bangalore, Karnataka, India

ISBN-13 (pbk): 978-1-4842-4946-8   ISBN-13 (electronic): 978-1-4842-4947-5
https://doi.org/10.1007/978-1-4842-4947-5

Managing Director, Apress Media LLC: Welmoed Spahr
Acquisitions Editor: Celestin Suresh John
Development Editor: James Markham
Coordinating Editor: Aditee Mirashi

Cover designed by eStudioCalamar

Cover image designed by Freepik (www.freepik.com)

Distributed to the book trade worldwide by Springer Science+Business Media New York, 233 Spring Street, 6th Floor, New York, NY 10013. Phone 1-800-SPRINGER, fax (201) 348-4505, e-mail orders-ny@springer-sbm.com, or visit www.springeronline.com. Apress Media, LLC is a California LLC and the sole member (owner) is Springer Science + Business Media Finance Inc (SSBM Finance Inc). SSBM Finance Inc is a **Delaware** corporation.

For information on translations, please e-mail rights@apress.com, or visit www.apress.com/rights-permissions.

Apress titles may be purchased in bulk for academic, corporate, or promotional use. eBook versions and licenses are also available for most titles. For more information, reference our Print and eBook Bulk Sales web page at www.apress.com/bulk-sales.

Any source code or other supplementary material referenced by the author in this book is available to readers on GitHub via the book's product page, located at www.apress.com/978-1-4842-4946-8. For more detailed information, please visit www.apress.com/source-code.

Printed on acid-free paper

Table of Contents

About the Author

Manohar Swamynathan is a data science practitioner and an avid programmer with over 14 years of experience in various data science related areas that include: data warehousing, business intelligence (BI), analytical tool development, ad-hoc analysis, predictive modeling, data science product development, consulting, formulating strategy and executing analytics programs. He's had a career covering the life cycle of data across different domains such as US mortgage banking, retail/e-commerce, insurance, and industrial IoT. He has a bachelor's degree with a specialization in physics, mathematics, and computers; and a master's degree in project management. He's currently living in Bengaluru, the silicon valley of India.

He's also involved in technical review of books on data science using Python and R.

About the Technical Reviewer

 **Jojo Moolayil** is an artificial intelligence professional and published author of three books on machine learning, deep learning, and IoT. He is currently working with Amazon Web Services as a Research Scientist – AI in their Vancouver, BC office.

He was born and raised in Pune, India and graduated from the University of Pune with a major in information technology engineering. His passion for problem solving and data-driven decision making led him to start a career with **Mu Sigma Inc.**, the world's largest pure play analytics provider. Here, he was responsible for developing machine learning and decision science solutions for large complex problems for healthcare and telecom giants. He later worked with **Flutura** (an IoT Analytics startup) and **General Electric** with a focus on industrial AI in Bangalore, India.

In his current role with AWS, he works on researching and developing large-scale AI solutions for combating fraud and enriching customers' payment experience in the cloud. He is also actively involved as a tech reviewer and AI consultant with leading publishers, and has reviewed more than a dozen books on machine learning, deep learning, and business analytics.

You can reach out to Jojo at

- www.jojomoolayil.com/
- www.linkedin.com/in/jojo62000
- https://twitter.com/jojo62000

Acknowledgments

I'm grateful to my mom, dad, and loving brother. I thank my wife Usha and son Jivin for providing me the space to write this book. I would like to express my gratitude to my colleagues/friends from current/previous organizations for their inputs, inspiration, and support. Thanks to Jojo for the encouragement to write this book and his technical review inputs. Big thanks to the Apress team for their constant support and help.

I would like to express my gratitude for the encouragement received from Ajit Jaokar.

Thanks for the input, feedback both positive and constructive provided by readers of the first edition of this book.

Finally, I would like to thank YOU, for showing an interest in this book and I sincerely hope to help your machine learning quest.

Introduction

This book is your practical guide to moving from novice to master in machine learning (ML) with Python 3 in six steps. The six steps path has been designed based on the "six degrees of separation" theory, which states that everyone and everything is a maximum of six steps away. Note that the theory deals with the *quality* of connections, rather than their existence. So a great effort has been taken to design an eminent yet simple six steps covering fundamentals to advanced topics gradually, to help a beginner walk his/her way from no or least knowledge of ML in Python all the way to becoming a master practitioner. This book is also helpful for current ML practitioners to learn advanced topics such as hyperparameter tuning, various ensemble techniques, natural language processing (NLP), deep learning, and the basics of reinforcement learning.

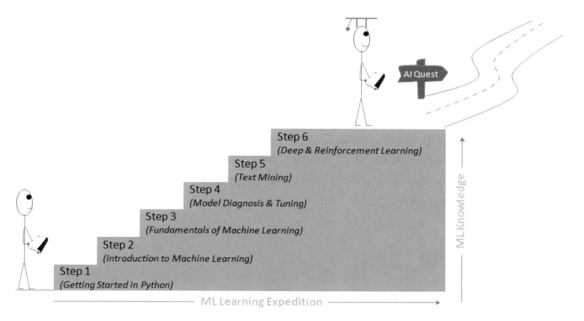

Figure 1. *Mastering machine learning with Python 3 in six steps*

Each topic has two parts: the first part will cover the theoretical concepts and the second part will cover practical implementation with different Python packages. The traditional approach of math to ML (i.e., learning all the mathematic then understanding how to implement them to solve problems) needs a great deal of time/effort, which has proved to be inefficient for working professionals looking to switch careers. Hence, the focus in this book has been more on simplification, such that the theory/math behind algorithms have been covered only to the extent required to get you started.

I recommend that you work with the book instead of reading it. Real learning goes on only through active participation. Hence, all the code presented in the book is available in the form of Jupyter notebooks to enable you to try out these examples yourselves and extend them to your advantage or interest as required later.

Who This Book Is For

This book will serve as a great resource for learning ML concepts and implementation techniques for:

- Python developers or data engineers looking to expand their knowledge or career into the machine learning area

- Current non-Python (R, SAS, SPSS, Matlab or any other language) ML practitioners looking to expand their implementation skills in Python

- Novice ML practitioners looking to learn advanced topics such as hyperparameter tuning, various ensemble techniques, natural language processing (NLP), deep learning, and the basics of reinforcement learning

What You Will Learn

Chapter 1, *Step 1: Getting Started in Python 3* will help you to set up the environment, and introduce you to the key concepts of Python 3 programming language relevant to machine learning. If you are already well versed in Python 3 basics, I recommend you to glance through the chapter quickly and move on to the next chapter.

Chapter 2, *Step 2: Introduction to Machine Learning*: Here you will learn about the history, evolution and different frameworks in practice for building ML systems.

I think this understanding is very important, as it will give you a broader perspective and set the stage for your further expedition. You'll understand the different types of ML (supervised/unsupervised/reinforcement learning). You will also learn the various concepts involved in core data analysis packages (NumPy, Pandas, Matplotlib) with example codes.

Chapter 3, *Step 3: Fundamentals of Machine Learning*: This chapter will expose you to various fundamental concepts involved in feature engineering, supervised learning (linear regression, nonlinear regression, logistic regression, time series forecasting, and classification algorithms), and unsupervised learning (clustering techniques, dimension reduction technique) with the help of Scikit-learn and statsmodel packages.

Chapter 4, *Step 4: Model Diagnosis and Tuning*: in this chapter you'll learn advanced topics around different model diagnosis, which covers the common problems that arise and various tuning techniques to overcome these issues to build efficient models. The topics include choosing the correct probability cutoff, handling an imbalanced dataset, the variance, and the bias issues. You'll also learn various tuning techniques such as ensemble models, and hyperparameter tuning using grid/random search.

Chapter 5, *Step 5: Text Mining and Recommender Systems*: Statistics say 70% of the data available in the business world is in the form of text, so text mining has vast scope across various domains. You will learn the building blocks and basic concepts to advanced NLP techniques. You'll also learn the recommender systems that are most commonly used to create personalization for customers.

Chapter 6, *Step 6: Deep and Reinforcement Learning*: There has been a great advancement in the area of artificial neural networks (ANNs) through deep learning techniques, and it has been the buzzword in recent times. You'll learn various aspects of deep learning such as multilayer perceptrons, convolutional neural networks (CNNs) for image classification, RNNs (recurrent neural network) for text classification, and transfer learning. You'll also use a Q-learning example to understand the concept of reinforcement learning.

Chapter 7, *Conclusion:* This chapter summarizes your six-step learning and includes quick tips that you should remember while starting with real-world machine learning problems.

Note   An appendix covering Generative Adversarial Networks (GAN) is available as part of this book's source code package, which can be accessed via the **Download Source Code** button located at `www.apress.com/9781484249468`.

## CHAPTER 1

Step 1: Getting Started in Python 3

In this chapter you will get a high-level overview about Python language and its core philosophy, how to set up the Python 3 development environment, and the key concepts around Python programming to get you started with basics. This chapter is an additional step or the prerequisite step for nonPython users. If you are already comfortable with Python, I would recommend you to quickly run through the contents to ensure you are aware of all the key concepts.

The Best Things in Life Are Free

It's been said that "*The best things in life are free!*" Python is an open source, high-level, object-oriented, interpreted, and general purpose dynamic programming language. It has a community-based development model. Its core design theory accentuates code readability, and its coding structure enables programmers to articulate computing concepts in fewer lines of code compared with other high-level programming languages such as Java, C, or C++.

The design philosophy of Python is well summarized by the document "The Zen of Python" (Python Enhancement Proposal, information entry number 20), which includes mottos such as:

- Beautiful is better than ugly—be consistent.

- Complex is better than complicated—use existing libraries.

- Simple is better than complex—keep it simple, stupid (KISS).

- Flat is better than nested—avoid nested ifs.

© Manohar Swamynathan 2019
M. Swamynathan, *Mastering Machine Learning with Python in Six Steps*,
https://doi.org/10.1007/978-1-4842-4947-5_1

- Explicit is better than implicit—be clear.

- Sparse is better than dense—separate code into modules.

- Readability counts—indent for easy readability.

- Special cases aren't special enough to break the rules—everything is an object.

- Errors should never pass silently—use good exception handling.

- Although practicality beats purity—if required, break the rules.

- Unless explicitly silenced—use error logging and traceability.

- In ambiguity, refuse the temptation to guess—Python syntax is simpler; however, many times we might take a longer time to decipher.

- Although the way may not be obvious at first—there is not only one way of achieving something.

- There should be, preferably, only one obvious way to do it—use existing libraries.

- If the implementation is hard to explain, it's a bad idea—if you can't explain in simple terms, then you don't understand it well enough.

- Now is better than never—there are quick/dirty ways to get the job done rather than trying too much to optimize.

- Although never is often better than right now—although there is a quick/dirty way, don't head on a path that will not allow a graceful way back.

- Namespaces are one honking great idea, so let's do more of those! Be specific.

- If the implementation is easy to explain, it may be a good idea—simplicity is good.

The Rising Star

Python was officially born on February 20, 1991, with version number 0.9.0. Its application cuts across various areas such as website development, mobile apps development, scientific and numeric computing, desktop GUI, and complex software development. Even though Python is a more general-purpose programming and scripting language, it has gained popularity over the past couple of years among data engineers, scientists, and Machine Learning (ML) enthusiasts.

There are well-designed development environments such as Jupyter Notebook and Spyder that allow for a quick examination of the data and enable developing of ML models interactively.

Powerful modules such as NumPy and Pandas exist for the efficient use of numeric data. Scientific computing is made easy with the SciPy package. A number of primary ML algorithms have been efficiently implemented in scikit-learn (also known as sklearn). HadooPy and PySpark provide a seamless work experience with big data technology stacks. Cython and Numba modules allow executing Python code on par with the speed of C code. Modules such as nosetest emphasize high quality, continuous integration tests, and automatic deployment.

Combining all of these has led many ML engineers to embrace Python as the choice of language to explore data, identify patterns, and build and deploy models to the production environment. Most importantly, the business-friendly licenses for various key Python packages are encouraging the collaboration of businesses and the open source community for the benefit of both worlds. Overall, the Python programming ecosystem allows for quick results and happy programmers. We have been seeing the trend of developers being part of the open source community to contribute to the bug fixes and new algorithms for use by the global community, at the same time protecting the core IP of the respective company they work for.

Choosing Python 2.x or Python 3.x

Python version 3.0, released in December 2008, is backward incompatible. That's because as there was big stress from the development team stressed separating binary data from textual data, and making all textual data automatically support Unicode so that project teams can work with multiple languages easily. As a result, any project migration from 2.x to 3.x required large changes. Python 2.x originally had a scheduled end-of-life (EOL) for 2015 but was extended for another 5 years to 2020.

Python 3 is a cutting edge, nicer and more consistent language. It is the future of the Python language and it fixes many of the problems that are present in Python 2. Table 1-1 shows some of the key differences.

Table 1-1. *Python 2 vs. Python 3*

Python 2	Python 3
It'll retire by 2020; till then it'll receive updates for security and bug fixes.	It has seen great adoption in the last two years; currently, 99.7% of key packages support Python 3.
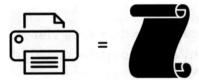 Print is a statement. Print "Hello World!"	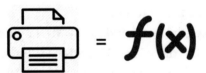 Print is a function. Print ("Hello World!")
ASCII Strings are by default stored as ASCII.	**UNICODE** Strings are by default stored as Unicode.
Rounds the integer division to the nearest whole number	Integer division returns the exact value without rounding to the nearest whole number.

As of now, Python 3 readiness (`http://py3readiness.org/`) shows that 360 of the 360 top packages for Python support 3.x. It is highly recommended that we use Python 3.x for development work.

I recommend Anaconda (Python distribution), BSD licensed, which gives you permission to use it commercially and for redistribution. It has around 474 packages, including the most important for most scientific applications, data analysis, and ML such as NumPy, SciPy, Pandas, Jupyter Notebook, matplotlib, and scikit-learn. It also provides a superior environment tool, conda, which allows you to easily switching between environments—even between Python 2 and 3 (if required). It is also updated very quickly as soon as a new version of a package is released; you can just do `conda update <packagename>` to update it.

You can download the latest version of Anaconda from their official website `https://www.anaconda.com/distribution/` and follow the installation instructions.

To install Python, refer to the following sections.

Windows

1. Download the installer, depending on your system configuration (32 or 64 bit).

2. Double-click the .exe file to install Anaconda and follow the installation wizard on your screen.

OSX

For Mac OS, you can install either through the graphical installer or from the command line.

Graphical Installer

1. Download the graphical installer.

2. Double-click the downloaded .pkg file and follow the installation wizard instructions on your screen.

Command Line Installer

1. Download the command-line installer

2. In your terminal window, type and follow the instructions: bash <Anaconda3-x.x.x-MacOSX-x86_64.sh>

Linux

1. Download the installer, depending on your system configuration.

2. In your terminal window, type and follow the instructions: bash Anaconda3-x.x.x-Linux-x86_xx.sh.

From Official Website

If you don't want to go with the Anaconda build pack, you can go to Python's official website `www.python.org/downloads/` and browse to appropriate OS section and download the installer. Note that OSX and most of the Linux comes with preinstalled Python, so there is no need for additional configuring.

When setting up a PATH for Windows, make sure to check the "Add Python to PATH option," when you run the installer. This will allow you to invoke Python interpreter from any directory.

If you miss ticking the "Add Python to PATH option", follow these steps:

1. Right click "My computer"

2. Click "Properties"

3. Click "Advanced system settings" in the side panel

4. Click "Environment Variables"

5. Click "New" below system variables.

6. In name, enter pythonexe (or anything you want).

7. In value, enter the path to your Python (example: C:\Python32\).

8. Now edit the Path variable (in the system part) and add %pythonexe%; to the end of what's already there.

Running Python

From the command line, type "Python" to open the interactive interpreter. A Python script can be executed at the command line using the syntax

```
python <scriptname.py>.
```

Key Concepts

There are many fundamental concepts in Python, and understanding them is essential for you to get started. The remainder of the chapter takes a concise look at them.

Python Identifiers

As the name suggests, identifiers help us to differentiate one entity from another. Python entities such as class, functions, and variables are called identifiers.

- It can be a combination of upper or lower case letters (a to z or A to Z).

- It can be any digits (0 to 9) or an underscore (_).

- The general rules to be followed for writing identifiers in Python:

    - It cannot start with a digit. For example, 1variable is not valid, whereas variable1 is valid.

    - Python reserved keywords (refer to Table 1-2) cannot be used as identifiers.

    - Except for underscore (_), special symbols like !, @, #, $, %, etc. cannot be part of the identifiers.

Keywords

Table 1-2 lists the set of reserved words used in Python to define the syntax and structure of the language. Keywords are case sensitive, and all the keywords are in lowercase except *True, False,* and *None*.

Table 1-2. *Python Keywords*

False	class	finally	Is	return
None	continue	for	lambda	try
True	Def	From	nonlocal	while
and	Del	Global	Not	with
as	elif	if	or	yield
assert	else	import	pass	
break	except	in	raise	

My First Python Program

Working with Python is comparatively a lot easier than other programming languages (Figure 1-1). Let's look at how an example of executing a simple print statement can be done in a single line of code. You can launch the Python interactive on the command prompt, type the following text, and press Enter.

```
>>> print ("Hello, Python World!")
```

Figure 1-1. *Python vs. others*

Code Blocks

It is very important to understand how to write code blocks in Python. Let's look at two key concepts around code blocks: indentations and suites.

Indentations

One of the most unique features of Python is its use of indentation to mark blocks of code. Each line of code must be indented by the same amount to denote a block of code in Python. Unlike most other programming languages, indentation is not used to help make the code look pretty. Indentation is required to indicate which block of code or statement belongs to current program structure (see Listings 1-1 and 1-2 for examples).

Suites

A collection of individual statements that makes a single code block are called suites in Python. A header line followed by a suite is required for compound or complex statements such as if, while, def, and class (we will understand each of these in detail in the later sections). Header lines begin with a keyword, and terminate with a colon (:) and are followed by one or more lines that make up the suite.

Listing 1-1. Example of Correct Indentation

```
# Correct indentation
print ("Programming is an important skill for Data Science")
print ("Statistics is an important skill for Data Science")
print ("Business domain knowledge is an important skill for Data Science")

# Correct indentation, note that if statement here is an example of suites
x = 1
if x == 1:
    print ('x has a value of 1')
else:
    print ('x does NOT have a value of 1')
```

Listing 1-2. Example of Incorrect Indentation

```
# incorrect indentation, program will generate a syntax error
# due to the space character inserted at the beginning of the second line
print ("Programming is an important skill for Data Science")
 print ("Statistics is an important skill for Data Science")
print ("Business domain knowledge is an important skill for Data Science")
3
# incorrect indentation, program will generate a syntax error
# due to the wrong indentation in the else statement
x = 1
if x == 1:
```

CHAPTER 1   STEP 1: GETTING STARTED IN PYTHON 3

```
    print ('x has a value of 1')
else:
 print ('x does NOT have a value of 1')
-------Output-----------
    print ("Statistics is an important skill for Data Science")
    ^

IndentationError: unexpected indent
```

Basic Object Types

Table 1-3 lists the Python object types. According to the Python data model reference, objects are Python's notion for data. All data in a Python program is represented by objects or by relations between objects. In a sense, and in conformance to Von Neumann's model of a "stored program computer," code is also represented by objects.

Every object has an identity, a type, and a value. Listing 1-3 provides example code to understand object types.

Table 1-3. *Python Object Types*

Type	Examples	Comments
None	None	# singleton null object
Boolean	True, False	
Integer	-1, 0, 1, sys.maxint	
Long	1L, 9787L	
Float	3.141592654	
	inf, float('inf')	# infinity
	-inf	# neg infinity
	nan, float('nan')	# not a number
Complex	2+8j	# note use of j
String	'this is a string', "also me"	# use single or double quote
	r'raw string', u'unicode string'	
Tuple	empty = ()	# empty tuple
	(1, True, 'ML')	# immutable list or unalterable list
List	empty = []	empty list
	[1, True, 'ML']	# mutable list or alterable list
Set	empty = set()	# empty set
	set(1, True, 'ML')	# mutable or alterable
dictionary	empty = {}	# mutable object or alterable object
	{'1':'A', '2':'AA', True = 1, False = 0}	
File	f = open('filename', 'rb')	

Listing 1-3. Code for Basic Object Types

```python
none = None              #singleton null object
boolean = bool(True)
integer = 1
Long = 3.14

# float
Float = 3.14
Float_inf = float('inf')
Float_nan = float('nan')

# complex object type, note the usage of letter j
Complex = 2+8j

# string can be enclosed in single or double quote
string = 'this is a string'
me_also_string = "also me"

List = [1, True, 'ML'] # Values can be changed

Tuple = (1, True, 'ML') # Values can not be changed

Set = set([1,2,2,2,3,4,5,5]) # Duplicates will not be stored

# Use a dictionary when you have a set of unique keys that map to values
Dictionary = {'a':'A', 2:'AA', True:1, False:0}

# lets print the object type and the value
print (type(none), none)
print (type(boolean), boolean)
print (type(integer), integer)
print (type(Long), Long)
print (type(Float), Float)
print (type(Float_inf), Float_inf)
print (type(Float_nan), Float_nan)
print (type(Complex), Complex)
print (type(string), string)
print (type(me_also_string), me_also_string)
print (type(Tuple), Tuple)
```

```
print (type(List), List)
print (type(Set), Set)
print (type(Dictionary), Dictionary)

----- output ------

<type 'NoneType'> None
<type 'bool'> True
<type 'int'> 1
<type 'float'> 3.14
<type 'float'> 3.14
<type 'float'> inf
<type 'float'> nan
<type 'complex'> (2+8j)
<type 'str'> this is a string
<type 'str'> also me
<type 'tuple'> (1, True, 'ML')
<type 'list'> [1, True, 'ML']
<type 'set'> set([1, 2, 3, 4, 5])
<type 'dict'> {'a': 'A', True: 1, 2: 'AA', False: 0}
```

When to Use List, Tuple, Set, or Dictionary

Four key, commonly used Python objects are list, tuple, set, and dictionary. It's important to understand when to use these, to be able to write efficient code.

- *List:* Use when you need an ordered sequence of homogenous collections whose values can be changed later in the program.

- *Tuple:* Use when you need an ordered sequence of heterogeneous collections whose values need not be changed later in the program.

- *Set:* It is ideal for use when you don't have to store duplicates and you are not concerned about the order of the items. You just want to know whether a particular value already exists or not.

- *Dictionary:* It is ideal for use when you need to relate values with keys, in order to look them up efficiently using a key.

Comments in Python

Single line comment: Any characters followed by the # (hash) and up to the end of the line are considered as part of the comment and the Python interpreter ignores them.

Multiline comments: Any characters between the strings """ (referred to as multiline string), that is, one at the beginning and end of your comments, will be ignored by the Python interpreter. Please refer to Listing 1-4 for a comments code example.

Listing 1-4. Example Code for Comments

```python
# This is a single line comment in Python
print("Hello Python World") # This is also a single line comment in Python

""" This is an example of a multi-line
the comment that runs into multiple lines.
Everything that is in between is considered as comments
"""
```

Multiline Statements

Python's oblique line continuation inside parentheses, brackets, and braces is the favorite way of casing longer lines. Using a backslash to indicate line continuation makes readability better; however, if needed you can add an extra pair of parentheses around the expression. It is important to indent the continued line of your code suitably. Note that the preferred place to break around the binary operator is after the operator, and not before it. Please refer to Listing 1-5 for Python code examples.

Listing 1-5. Example Code for Multiline Statements

```python
# Example of implicit line continuation
x = ('1' + '2' +
    '3' + '4')

# Example of explicit line continuation
y = '1' + '2' + \
    '11' + '12'

weekdays = ['Monday', 'Tuesday', 'Wednesday',
'Thursday', 'Friday']
```

```
weekend = {'Saturday',
           'Sunday'}

print ('x has a value of', x)
print ('y has a value of', y)
print (weekdays)
print (weekend)

------ output -------
('x has a value of', '1234')
('y has a value of', '1234')
['Monday', 'Tuesday', 'Wednesday', 'Thursday', 'Friday']
set(['Sunday', 'Saturday'])
```

Multiple Statements on a Single Line

Python also allows multiple statements on a single line through the usage of the semicolon (;), given that the statement does not start a new code block. Listing 1-6 provides a code example.

Listing 1-6. Code Example for Multiple Statements on a Single Line

```
import os; x = 'Hello'; print (x)
```

Basic Operators

In Python, operators are the special symbols that can manipulate the value of operands. For example, let's consider the expression $1 + 2 = 3$. Here, 1 and 2 are called operands, which are the value on which operators operate, and the symbol + is called operator.

Python language supports the following types of operators:

- Arithmetic operators

- Comparison or Relational operators

- Assignment operators

- Bitwise operators

- Logical operators

- Membership operators

- Identity operators

Let's learn all operators through examples, one by one.

Arithmetic Operators

Arithmetic operators (listed in Table 1-4) are useful for performing mathematical operations on numbers such as addition, subtraction, multiplication, division, etc. Please refer to Listing 1-7 for a code example.

Table 1-4. *Arithmetic Operators*

Operator	Description	Example
+	Addition	x + y = 30
-	Subtraction	x − y = -10
*	Multiplication	x * y = 200
/	Division	y / x = 2
%	Modulus	y % x = 0
** Exponent	Exponentiation	x**b = 10 to the power 20
//	Floor Division - Integer division rounded toward minus infinity	9//2 = 4 and 9.0//2.0 = 4.0, -11//3 = -4, -11.0/

Listing 1-7. Example Code for Arithmetic Operators

```
# Variable x holds 10 and variable y holds 5
x = 10
y = 5

# Addition
print ("Addition, x(10) + y(5) = ", x + y)

# Subtraction
print ("Subtraction, x(10) - y(5) = ", x - y)

# Multiplication
print ("Multiplication, x(10) * y(5) = ", x * y)
```

```
# Division
print ("Division, x(10) / y(5) = ",x / y)

# Modulus
print ("Modulus, x(10) % y(5) = ", x % y)

# Exponent
print ("Exponent, x(10)**y(5) = ", x**y)

# Integer division rounded towards minus infinity
print ("Floor Division, x(10)//y(5) = ", x//y)

-------- output --------

Addition, x(10) + y(5) =   15
Subtraction, x(10) - y(5) =  5
Multiplication, x(10) * y(5) =   50
Divions, x(10) / y(5) =   2.0
Modulus, x(10) % y(5) =   0
Exponent, x(10)**y(5) =   100000
Floor Division, x(10)//y(5) =   2
```

Comparison or Relational Operators

As the name suggests, the comparison or relational operators listed in Table 1-5 are useful to compare values. They would return True or False as a result for a given condition. Refer to Listing 1-8 for code examples.

Table 1-5. *Comparison or Relational Operators*

Operator	Description	Example
==	The condition becomes True, if the values of two operands are equal.	(10 == 5) is not True.
!=	The condition becomes True, if the values of two operands are not equal.	(10 != 5) is True.
>	The condition becomes True, if the value of left operand is greater than the value of right operand.	(10 > 5) is not True.
<	The condition becomes True, if the value of left operand is less than the value of right operand.	(10 < 5) is True.
>=	The condition becomes True, if the value of left operand is greater than or equal to the value of right operand.	(10 >= 5) is not True.
<=	The condition becomes True, if the value of left operand is less than or equal to the value of right operand.	(10 <= 5) is True.

Listing 1-8. Example Code for Comparision/Relational Operators

```
# Variable x holds 10 and variable y holds 5
x = 10
y = 5

# Equal check operation
print ("Equal check, x(10) == y(5) ", x == y)

# Not Equal check operation
print ("Not Equal check, x(10) != y(5) ", x != y)

# Less than check operation
print ("Less than check, x(10) <y(5) ", x<y)

# Greater check operation
print ("Greater than check, x(10) >y(5) ", x>y)

# Less than or equal check operation
print ("Less than or equal to check, x(10) <= y(5) ", x<= y)

# Greater than or equal to check operation
print ("Greater than or equal to check, x(10) >= y(5) ", x>= y)
```

```
-------- output --------
Equal check, x(10) == y(5)  False
Not Equal check, x(10) != y(5)  True
Less than check, x(10) <y(5)  False
Greater than check, x(10) >y(5)  True
Less than or equal to check, x(10) <= y(5)  False
Greater than or equal to check, x(10) >= y(5)  True
```

Assignment Operators

In Python, assignment operators listed in Table 1-6 are used for assigning values to variables. For example, consider x = 5; it is a simple assignment operator that assigns the numeric value 5, which is on the right side of the operator, to the variable x on the left side. There is a range of compound operators in Python like x += 5 that adds to the variable and later assigns the same. It is as good as x = x + 5. Refer to Listing 1-9 for code examples.

Table 1-6. *Assignment Operators*

Operator	Description	Example
=	Assigns values from right side operands to left side operand	z = x + y assigns value of x + y to z
+= Add AND	It adds right operand to the left operand and assigns the result to left operand	z += x is equivalent to z = z + x
-= Subtract AND	It subtracts right operand from the left operand and assigns the result to left operand.	z -= x is equivalent to z = z - x
*= Multiply AND	It multiplies right operand with the left operand and assigns the result to left operand.	z *= x is equivalent to z = z * x
/= Divide AND	It divides left operand with the right operand and assigns the result to left operand.	z /= x is equivalent to z = z/ xz/= x is equivalent to z = z / x
%= Modulus AND	It takes modulus using two operands and assigns the result to left operand.	z %= x is equivalent to z = z % x
**= Exponent AND	It performs exponential (power) calculation on operators and assigns value to the left operand.	z **= x is equivalent to z = z** x
//= Floor Division	It performs floor division on operators and assigns value to the left operand.	z //= x is equivalent to z = z// x

Listing 1-9. Example Code for Assignment Operators

```python
# Variable x holds 10 and variable y holds 5
x = 5
y = 10

x += y
print ("Value of a post x+=y is ", x)

x *= y
print ("Value of a post x*=y is ", x)

x /= y
print ("Value of a post x/=y is ", x)

x %= y
print ("Value of a post x%=y is ", x)

x **= y
print ("Value of x post x**=y is ", x)

x //= y
print ("Value of a post x//=y is ", x)
-------- output --------
Value of a post x+=y is   15
Value of a post x*=y is   150
Value of a post x/=y is   15.0
Value of a post x%=y is   5.0
Value of x post x**=y is   9765625.0
Value of a post x//=y is   976562.0
```

Bitwise Operators

As you might be aware, everything in a computer is represented by bits, that is, a series of 0s (zero) and 1s (one). Bitwise operators listed in Table 1-7 enable us to directly operate or manipulate bits. Let's understand the basic bitwise operations. One of the key usages of bitwise operators is for parsing hexadecimal colors.

Bitwise operators are known to be confusing for newbies to Python programming, so don't be anxious if you don't understand usability at first. The fact is that you aren't really going to see bitwise operators in your everyday ML programming. However, it is good to be aware of these operators.

For example, let's assume that x = 10 (in binary 0000 1010) and y = 4 (in binary 0000 0100). Refer to Listing 1-10 for code examples.

Table 1-7. *Bitwise Operators*

Operator	Description	Example
& Binary AND	This operator copies a bit to the result if it exists in both operands.	(x&y) (means 0000 0000)
| Binary OR	This operator copies a bit if it exists in either operand.	(x | y) = 14 (means 0000 1110)
^ Binary XOR	This operator copies the bit if it is set in one operand but not both.	(x ^ y) = 14 (means 0000 1110)
~ Binary Ones Complement	This operator is unary and has the effect of "flipping" bits.	(~x) = -11 (means 1111 0101)
<< Binary Left Shift	The left operands value is moved left by the number of bits specified by the right operand.	x<< 2= 42 (means 0010 1000)
>> Binary Right Shift	The left operands value is moved right by the number of bits specified by the right operand.	x>> 2 = 2 (means 0000 0010)

Listing 1-10. Example Code for Bitwise Operators

```
# Basic six bitwise operations
# Let x = 10 (0000 1010 in binary) and y = 4 (0000 0100 in binary)
x = 10
y = 4

print (x >> y)  # Right Shift
print (x << y)  # Left Shift
print (x & y)   # Bitwise AND
print (x | y)   # Bitwise OR
print (x ^ y) # Bitwise XOR
```

```
print (~x)      # Bitwise NOT
-------- output --------
0
160
0
14
14
-11
```

Logical Operators

The AND, OR, NOT operators are called logical operators and listed in Table 1-8. These are useful to check two variables against a given condition, and the result will be True or False appropriately. Refer to Listing 1-11 for code examples.

Table 1-8. *Logical Operators*

Operator	Description	Example
and Logical AND	If both the operands are True, then condition becomes True.	(var1 and var2) is True.
or Logical OR	If any of the two operands are non-zero, then condition becomes True.	(var1 or var2) is True.
not Logical NOT	Used to reverse the logical state of its operand	Not (var1 and var2) is False.

Listing 1-11. Example Code for Logical Operators

```
var1 = True
var2 = False
print('var1 and var2 is',var1 and var2)
print('var1 or var2 is',var1 or var2)
print('not var1 is',not var1)
```

```
-------- output --------

var1 and var2 is False
var1 or var2 is True
not var1 is False
```

Membership Operators

Membership operators listed in Table 1-9 are useful to test if a value is found in a sequence, that is, string, list, tuple, set, or dictionary. There are two membership operators in Python: "in" and "not in." Note that we can only test for the presence of a key (and not the value) in the case of a dictionary. Refer to Listing 1-12 for code examples.

Table 1-9. *Membership Operators*

Operator	Description	Example
in	Results to True if a value is in the specified sequence, and False otherwise	var1 in var2
not in	Results to True if a value is not in the specified sequence, and False otherwise	var1 not in var2

Listing 1-12. Example Code for Membership Operators

```
var1 = 'Hello world'      # string
var2 = {1:'a',2:'b'}      # dictionary
print('H' in var1)
print('hello' not in var1)
print(1 in var2)
print('a' in var2)
-------- output --------
True
True
True
False
```

Identity Operators

Identity operators listed in Table 1-10 are useful to test if two variables are present on the same part of the memory. There are two identity operators in Python: "*is*" and "*is not*." Note that two variables having equal values do not imply they are identical. Refer to Listing 1-13 for code examples.

Table 1-10. *Identity Operators*

Operator	Description	Example
is	Results to True if the variables on either side of the operator point to the same object, and False otherwise	var1 is var2
is not	Results to False if the variables on either side of the operator point to the same object, and True otherwise	Var1 is not var2

Listing 1-13. Example Code for Identity Operators

```
var1 = 5
var1 = 5
var2 = 'Hello'
var2 = 'Hello'
var3 = [1,2,3]
var3 = [1,2,3]
print(var1 is not var1)
print(var2 is var2)
print(var3 is var3)
-------- output --------
False
True
False
```

Control Structures

A control structure is the fundamental choice or decision-making process in programming. It is a chunk of code that analyzes values of variables and decides a direction to go based on a given condition. In Python there are mainly two types of control structures: selections and iterations.

Selections

Selections statements allow programmers to check a condition and, based on the result, perform different actions. There are two versions of this useful construct: 1) if and 2) if...else. Refer to Listings 1-14 through 1-16 for code examples.

Listing 1-14. Example Code for a Simple "if" Statement

```
var = -1
if var < 0:
    print (var)
    print ("the value of var is negative")

# If the suite of an if clause consists only of a single line, it may go on
the same line as the header statement
if ( var  == -1 ) : print ("the value of var is negative")
-------- output --------
-1
the value of var is negative
the value of var is negative
```

Listing 1-15. Example Code for "if else" Statement

```
var = 1
if var < 0:
    print ("the value of var is negative")
    print (var)
else:
    print ("the value of var is positive")
    print (var)
-------- output --------
the value of var is positive
1
```

Listing 1-16. Example Code for Nested "if else" Statements

```
score = 95

if score >= 99:
    print('A')
elif score >=75:
    print('B')
elif score >= 60:
    print('C')
elif score >= 35:
    print('D')
else:
    print('F')
-------- output --------
B
```

Iterations

A loop control statement enables us to execute a single or a set of programming statements multiple times until a given condition is satisfied. Python provides two essential looping statements: 1) "for" and 2) "while."

For loop: It allows us to execute a code block for a specific number of times or against a specific condition until it is satisfied. Refer to Listings 1-17 through 1-19 for code examples.

Listing 1-17. Example Code for a "for" Loop Statement

```
# First Example
print ("First Example")
for item in [1,2,3,4,5]:
    print ('item :', item)

# Second Example
print ("Second Example")
letters = ['A', 'B', 'C']
for letter in letters:
    print ('First loop letter :', letter)
```

```
# Third Example - Iterating by sequency index
print ("Third Example")
for index in range(len(letters)):
    print ('First loop letter :', letters[index])

# Fourth Example - Using else statement
print ("Fourth Example")
for item in [1,2,3,4,5]:
    print ('item :', item)
else:
    print ('looping over item complete!')
----- output ------
First Example
item : 1
item : 2
item : 3
item : 4
item : 5
Second Example
First loop letter : A
First loop letter : B
First loop letter : C
Third Example
First loop letter : A
First loop letter : B
First loop letter : C
Fourth Example
item : 1
item : 2
item : 3
item : 4
item : 5
looping over item complete!
```

While loop: The while statement repeats a set of code until the condition is true.

Listing 1-18. Example Code for a "while" Loop Statement

```
count = 0
while (count < 5):
    print ('The count is:', count)
    count = count + 1
----- output ------
The count is: 0
The count is: 1
The count is: 2
The count is: 3
The count is: 4
```

Caution   If a condition never becomes FALSE, the loop becomes an infinite loop.

An else statement can be used with a while loop and the else will be executed when the condition becomes false.

Listing 1-19. Example Code for a "while" with an "else" Statement

```
count = 0
while count < 5:
    print (count, " is  less than 5")
    count = count + 1
else:
    print (count, " is not less than 5")
----- output ------
0  is  less than 5
1  is  less than 5
2  is  less than 5
3  is  less than 5
4  is  less than 5
5  is not less than 5
```

Lists

Python's lists are the most flexible data type. They can be created by writing a list of comma separated values between square brackets. Note that that the items in the list need not be of the same data type. Table 1-11 summarizes list operations; refer to Listings 1-20 through 1-24 for code examples.

Table 1-11. *Python List Operations*

Description	Python Expression	Example	Results
Creating a list of items	[item1, item2, …]	list = ['a','b','c','d']	['a','b','c','d']
Accessing items in list	list[index]	list = ['a','b','c','d'] list[2]	c
Length	len(list)	len([1, 2, 3])	3
Concatenation	list_1 + list_2	[1, 2, 3] + [4, 5, 6]	[1, 2, 3, 4, 5, 6]
Repetition	list $*$ int	['Hello'] $*$ 3	['Hello', 'Hello', 'Hello']
Membership	item in list	3 in [1, 2, 3]	TRUE
Iteration	for x in list: print(x)	for x in [1, 2, 3]: print(x)	1 2 3
Count from the right	list[-index]	list = [1,2,3]; list[-2]	2
Slicing fetches sections	list[index:]	list = [1,2,3]; list[1:]	[2,3]
Return max item	max(list)	max([1,2,3,4,5])	5
Return min item	min(list)	max([1,2,3,4,5])	1
Append object to list	list.append(obj)	[1,2,3,4].append(5)	[1,2,3,4,5]
Count item occurrence	list.count(obj)	[1,1,2,3,4].count(1)	2
Append content of sequence to list	list.extend(seq)	['a', 1].extend(['b', 2])	['a', 1, 'b', 2]
Return the first index position of item	list.index(obj)	['a', 'b','c',1,2,3].index('c')	2

(continued)

Table 1-11. (*continued*)

Description	Python Expression	Example	Results
Insert object to list at a desired index	list.insert(index, obj)	['a', 'b','c',1,2,3].insert(4, 'd')	['a', 'b','c','d', 1,2,3]
Remove and return last object from list	list.pop(obj=list[-1])	['a', 'b','c',1,2,3].pop() ['a', 'b','c',1,2,3].pop(2)	3c
Remove object from list	list.remove(obj)	['a', 'b','c',1,2,3].remove('c')	['a', 'b', 1,2,3]
Reverse objects of list in place	list.reverse()	['a', 'b','c',1,2,3].reverse()	[3,2,1,'c','b','a']
Sort objects of list	list.sort()	['a', 'b','c',1,2,3].sort() ['a', 'b','c',1,2,3].sort (reverse = True)	[1,2,3,'a', 'b','c'] ['c','b','a',3,2,1]

Listing 1-20. Example Code for Accessing Lists

```
list_1 = ['Statistics', 'Programming', 2016, 2017, 2018];
list_2 = ['a', 'b', 1, 2, 3, 4, 5, 6, 7 ];

# Accessing values in lists
print ("list_1[0]: ", list_1[0])
print ("list2_[1:5]: ", list_2[1:5])
---- output ----

list_1[0]:  Statistics
list2_[1:5]:  ['b', 1, 2, 3]
```

Listing 1-21. Example Code for Adding New Values to Lists

```
print ("list_1 values: ", list_1)
list_1.append(2019)
print ("list_1 values post append: ", list_1)
---- output ----
list_1 values:  ['Statistics', 'Programming', 2016, 2017, 2018]
list_1 values post append:  ['Statistics', 'Programming', 2016, 2017,
2018, 2019]
```

Listing 1-22. Example Code for Updating Existing Values of Lists

```
# Updating existing values of list
print ("Value available at index 2 : ", list_1[2])
list_1[2] = 2015;
print ("New value available at index 2 : ", list_1[2])
---- output ----
Values of list_1:  ['Statistics', 'Programming', 2016, 2017, 2018, 2019]
Value available at index 2 :   2016
New value available at index 2 :   2015
```

Listing 1-23. Example Code for Deleting a List Element

```
# Deleting list elements
print ("list_1 values: ", list_1)
del list_1[5];
print ("After deleting value at index 2 : ", list_1)
---- output ----
list_1 values:  ['Statistics', 'Programming', 2015, 2017, 2018, 2019]
After deleting value at index 2 :  ['Statistics', 'Programming', 2015,
2017, 2018]
```

Listing 1-24. Example Code for Basic Operations on Lists

```
# Basic Operations
print ("Length: ", len(list_1))
print ("Concatenation: ", [1,2,3] + [4, 5, 6])
print ("Repetition :", ['Hello'] * 4)
print ("Membership :", 3 in [1,2,3])
print ("Iteration :" )
for x in [1,2,3]: print (x)

# Negative sign will count from the right
print ("slicing :", list_1[-2])
# If you don't specify the end explicitly, all elements from the specified
start index will be printed
print ("slicing range: ", list_1[1:])
```

```python
# Comparing elements of lists
# cmp function is only available in Python 2 and not 3, so if you still
need it you could use the below custom function
def cmp(a, b):
    return (a > b) - (a < b)

print ("Compare two lists: ", cmp([1,2,3, 4], [1,2,3]))
print ("Max of list: ", max([1,2,3,4,5]))
print ("Min of list: ", min([1,2,3,4,5]))
print ("Count number of 1 in list: ", [1,1,2,3,4,5,].count(1))
list_1.extend(list_2)
print ("Extended :", list_1)
print ("Index for Programming : ", list_1.index( 'Programming'))
print (list_1)
print ("pop last item in list: ", list_1.pop())
print ("pop the item with index 2: ", list_1.pop(2))
list_1.remove('b')
print ("removed b from list: ", list_1)
list_1.reverse()
print ("Reverse: ", list_1)
list_1 = ['a','b','c']
list_1.sort()
print ("Sort ascending: ", list_1)
list_1.sort(reverse = True)
print ("Sort descending: ", list_1)
---- output ----

Length:  5
Concatenation:  [1, 2, 3, 4, 5, 6]
Repetition : ['Hello', 'Hello', 'Hello', 'Hello']
Membership : True
Iteration :
1
2
3
slicing : 2017
```

```
slicing range:  ['Programming', 2015, 2017, 2018]
Compare two lists:  1
Max of list:  5
Min of list:  1
Count number of 1 in list:  2
Extended : ['Statistics', 'Programming', 2015, 2017, 2018, 'a', 'b', 1, 2,
3, 4, 5, 6, 7]
Index for Programming :  1
['Statistics', 'Programming', 2015, 2017, 2018, 'a', 'b', 1, 2, 3, 4, 5, 6, 7]
pop last item in list:  7
pop the item with index 2:  2015
removed b from list:  ['Statistics', 'Programming', 2017, 2018, 'a', 1, 2,
3, 4, 5, 6]
Reverse:  [6, 5, 4, 3, 2, 1, 'a', 2018, 2017, 'Programming', 'Statistics']
Sort ascending:  ['a', 'b', 'c']
Sort descending:  ['c', 'b', 'a']
```

Tuples

A Python tuple is a sequence or series of immutable Python objects very much similar to the lists. However, there exist some essential differences between lists and tuples:

1. Unlike a list, the objects of a tuple cannot be changed.

2. A tuple is defined by using parentheses, but lists are defined by square brackets.

Table 1-12 summarizes tuple operations; refer to Listings 1-25 to 1-28 for code examples.

Table 1-12. *Python Tuple Operations*

Description	Python Expression	Example	Results
Creating a tuple	(item1, item2, …) () # empty tuple (item1,)# Tuple with one item, note comma is required	tuple = ('a','b','c','d',1,2,3) tuple = () tuple = (1,)	('a','b','c','d', 1,2,3) () 1
Accessing items in a tuple	tuple[index] tuple[start_index:end_index]	tuple = ('a','b','c','d',1,2,3) tuple[2] tuple[0:2]	c a, b, c
Deleting a tuple	del tuple_name	del tuple	
Length	len(tuple)	len((1, 2, 3))	3
Concatenation	tuple_1 + tuple_2	(1, 2, 3) + (4, 5, 6)	(1, 2, 3, 4, 5, 6)
Repetition	tuple * int	('Hello',) * 4	('Hello', 'Hello', 'Hello', 'Hello')
Membership	item in tuple	3 in (1, 2, 3)	TRUE
Iteration	for x in tuple: print(x)	for x in (1, 2, 3): print(x)	1 2 3
Count from the right	tuple[-index]	tuple = (1,2,3); list[-2]	2
Slicing fetches sections	tuple[index:]	tuple = (1,2,3); list[1:]	(2,3)
Return max item	max(tuple)	max((1,2,3,4,5))	5
Return min item	min(tuple)	max((1,2,3,4,5))	1
Convert a list to tuple	tuple(seq)	tuple([1,2,3,4])	(1,2,3,4,5)

Listing 1-25. Example Code for Creating a Tuple

```
Tuple = ( )
print ("Empty Tuple: ", Tuple)

Tuple = (1,)
print ("Tuple with a single item: ", Tuple)
```

```
Tuple = ('a','b','c','d',1,2,3)
print ("Sample Tuple :", Tuple)
---- output ----
Empty Tuple:  ()
Tuple with a single item:  (1,)
Sample Tuple : ('a', 'b', 'c', 'd', 1, 2, 3)
```

Listing 1-26. Example Code for Accessing a Tuple

```
Tuple = ('a', 'b', 'c', 'd', 1, 2, 3)

print ("3rd item of Tuple:", Tuple[2])
print ("First 3 items of Tuple", Tuple[0:2])
---- output ----
3rd item of Tuple: c
First 3 items of Tuple ('a', 'b')
```

Listing 1-27. Example Code for Deleting a Tuple

```
print ("Sample Tuple: ",Tuple)
del Tuple
print (Tuple) # Will throw an error message as the tuple does not exist

---- output ----
Sample Tuple:  ('a', 'b', 'c', 'd', 1, 2, 3)
---------------------------------------------------------------------------
Sample Tuple:  ('a', 'b', 'c', 'd', 1, 2, 3)
---------------------------------------------------------------------------
NameError                                 Traceback (most recent call last)
<ipython-input-6-002eefa7c22f> in <module>
      4 print ("Sample Tuple: ",Tuple)
      5 del Tuple
----> 6 print (Tuple) # Will throw an error message as the tuple does not exist

NameError: name 'Tuple' is not defined
```

Listing 1-28. Example Code for Basic Tuple Operations (Not Exhaustive)

```
# Basic Tuple operations
Tuple = ('a','b','c','d',1,2,3)

print ("Length of Tuple: ", len(Tuple))
Tuple_Concat = Tuple + (7,8,9)
print ("Concatinated Tuple: ", Tuple_Concat)

print ("Repetition: ", (1, 'a',2, 'b') * 3)
print ("Membership check: ", 3 in (1,2,3))

# Iteration
for x in (1, 2, 3): print (x)

print ("Negative sign will retrieve item from right: ", Tuple_Concat[-2])
print ("Sliced Tuple [2:] ", Tuple_Concat[2:])

# Find max
print ("Max of the Tuple (1,2,3,4,5,6,7,8,9,10): ",
max((1,2,3,4,5,6,7,8,9,10)))
print ("Min of the Tuple (1,2,3,4,5,6,7,8,9,10): ",
min((1,2,3,4,5,6,7,8,9,10)))
print ("List [1,2,3,4] converted to tuple: ", type(tuple([1,2,3,4])))
---- output ----
Length of Tuple:  7
Concatinated Tuple:  ('a', 'b', 'c', 'd', 1, 2, 3, 7, 8, 9)
Repetition:  (1, 'a', 2, 'b', 1, 'a', 2, 'b', 1, 'a', 2, 'b')
Membership check:  True
1
2
3
Negative sign will retrieve an item from right:  8
Sliced Tuple [2:]  ('c', 'd', 1, 2, 3, 7, 8, 9)
Max of the Tuple (1,2,3,4,5,6,7,8,9,10):  10
Min of the Tuple (1,2,3,4,5,6,7,8,9,10):  1
List [1,2,3,4] converted to tuple:  <type 'tuple'>
```

Sets

As the name implies, sets are the implementations of mathematics sets, whose key characteristics are as follows:

- The collection of items is unordered.

- No duplicate items will be stored, which means each item is unique.

- Sets are mutable, which means the items in a set can be changed.

An item can be added or removed from sets. Mathematical set operations such as union, intersection, etc. can be performed on Python sets. Table 1-13 summarizes Python set operations, Listing 1-29 shows example code for creating sets, and Listing 1-30 shows example code for accessing set elements.

Table 1-13. *Python Set Operations*

Description	Python Expression	Example	Results
Creating a set	set{item1, item2, ...} set() # empty set	languages = set(['Python', 'R', 'SAS', 'Julia'])	set(['SAS', 'Python', 'R', 'Julia'])
Adding an item/element to a set	add()	languages.add('SPSS')	set(['SAS', 'SPSS', 'Python', 'R', 'Julia'])
Removing all items/ elements from a set	clear()	languages.clear()	set([])
Returning a copy of a set	copy()	lang = languages.copy() print(lang)	set(['SAS', 'SPSS', 'Python', 'R', 'Julia'])
Removing an item/ element from set if it is a member. (Do nothing if the element is not in set)	discard()	languages = set(['C', 'Java', 'Python', 'Data Science', 'Julia', 'SPSS', 'AI', 'R', 'SAS', 'Machine Learning']) languages.discard('AI')	set(['C', 'Java', 'Python', 'Data Science', 'Julia', 'SPSS', 'R', 'SAS', 'Machine Learning'])
Removing an item/ element from a set. It the element is not a member, raise a KeyError.	remove()	languages = set(['C', 'Java', 'Python', 'Data Science', 'Julia', 'SPSS', 'AI', 'R', 'SAS', 'Machine Learning']) languages.remove('AI')	set(['C', 'Java', 'Python', 'Data Science', 'Julia', 'SPSS', 'R', 'SAS', 'Machine Learning'])

(continued)

Table 1-13. (*continued*)

Description	Python Expression	Example	Results
Removing and returning an arbitrary set element. Raise a KeyError if the set is empty.	pop()	languages = set(['C', 'Java', 'Python', 'Data Science', 'Julia', 'SPSS', 'AI', 'R', 'SAS', 'Machine Learning']) print("Removed:", (languages. pop())) print(languages)	Removed: Cset (['Java', 'Python', 'Data Science', 'Julia', 'SPSS', 'R', 'SAS', 'Machine Learning'])
Returning the difference of two or more sets as a new set	difference()	# initialize A and B A = {1, 2, 3, 4, 5} B = {4, 5, 6, 7, 8} A.difference(B)	{1, 2, 3}
Removing all item/ element of another set from this set	difference_update()	# initialize A and B A = {1, 2, 3, 4, 5} B = {4, 5, 6, 7, 8} A.difference_update(B) print(A)	set([1, 2, 3])
Returning the intersection of two sets as a new set	intersection()	# initialize A and B A = {1, 2, 3, 4, 5} B = {4, 5, 6, 7, 8} A.intersection(B)	{4, 5}
Updating the set with the intersection of itself and another	intersection_update()	# initialize A and B A = {1, 2, 3, 4, 5} B = {4, 5, 6, 7, 8} A.intersection_update(B) print(A)	set([4, 5])
Returning True if two sets have a null intersection	isdisjoint()	# initialize A and B A = {1, 2, 3, 4, 5} B = {4, 5, 6, 7, 8} A.isdisjoint(B)	FALSE

(*continued*)

Table 1-13. (*continued*)

Description	Python Expression	Example	Results

Returning True if another set contains this set	issubset()	# initialize A and B
		A = {1, 2, 3, 4, 5}
		B = {4, 5, 6, 7, 8}
		print (A.issubset(B))
			FALSE

Returning True if this set contains another set	issuperset()	# initialize A and B
		A = {1, 2, 3, 4, 5}
		B = {4, 5, 6, 7, 8}
		print (A.issuperset(B))
			FALSE

Returning the symmetric difference of two sets as a new set	symmetric_ difference()	# initialize A and B
		A = {1, 2, 3, 4, 5}
		B = {4, 5, 6, 7, 8}
		A.symmetric_difference(B)
			{1, 2, 3, 6, 7, 8}

Updating a set with the symmetric difference of itself and another	symmetric_ difference_update()	# initialize A and B
		A = {1, 2, 3, 4, 5}
		B = {4, 5, 6, 7, 8}
		A.symmetric_difference(B)
		print(A)
		A.symmetric_difference_
		update(B)
		print(A)
			set([1, 2, 3, 6, 7, 8])

Returning the union of sets in a new set	union()	# initialize A and B
		A = {1, 2, 3, 4, 5}
		B = {4, 5, 6, 7, 8}
		A.union(B)
		print(A)
			set([1, 2, 3, 4, 5])

Updating a set with the union of itself and others	update()	# initialize A and B
		A = {1, 2, 3, 4, 5}
		B = {4, 5, 6, 7, 8}
		A.update(B)
		print(A)
			set([1, 2, 3, 4, 5, 6, 7, 8])

(*continued*)

Table 1-13. (*continued*)

Description	Python Expression	Example	Results
Returning the length (the number of items) in the set	len()	A = {1, 2, 3, 4, 5} len(A)	5
Returning the largest item in the set	max()	A = {1, 2, 3, 4, 5} max(A)	1
Returning the smallest item in the set	min()	A = {1, 2, 3, 4, 5} min(A)	5
Returning a new sorted list from elements in the set. Does not sort the set	sorted()	A = {1, 2, 3, 4, 5} sorted(A)	[4, 5, 6, 7, 8]
Returning the sum of all item/element in the set	sum()	A = {1, 2, 3, 4, 5} sum(A)	15

Listing 1-29. Example Code for Creating Sets

```
# Creating an empty set
languages = set()
print (type(languages), languages)

languages = {'Python', 'R', 'SAS', 'Julia'}
print (type(languages), languages)

# set of mixed datatypes
mixed_set = {"Python", (2.7, 3.4)}
print (type(mixed_set), languages)
---- output ----
<class 'set'> set()
<class 'set'> {'R', 'Python', 'SAS', 'Julia'}
<class 'set'> {'R', 'Python', 'SAS', 'Julia'}
```

Listing 1-30. Example Code for Accessing Set Elements

```
print (list(languages)[0])
print (list(languages)[0:3])
```

```
---- output ----
R
['R', 'Python', 'SAS']
```

Changing Sets in Python

Although sets are mutable, indexing on them will have no meaning due to the fact that they are unordered. So sets do not support accessing or changing an item/element using indexing or slicing. The add() method can be used to add a single element, and the update() method for adding multiple elements. Note that the update() method can take the argument in the format of a tuple, lists, strings, or other sets. However, in all cases, the duplicates are ignored. Refer to Listing 1-31 for code examples for changing a set elements.

Listing 1-31. Example Code for Changing Set Elements

```
# initialize a set
languages = {'Python', 'R'}
print(languages)

# add an element
languages.add('SAS')
print(languages)

# add multiple elements
languages.update(['Julia','SPSS'])
print(languages)

# add list and set
languages.update(['Java','C'], {'Machine Learning','Data Science','AI'})
print(languages)
---- output ----
{'R', 'Python'}
{'R', 'Python', 'SAS'}
{'Julia', 'R', 'Python', 'SAS', 'SPSS'}
{'Julia', 'Machine Learning', 'R', 'Python', 'SAS', 'Java', 'C', 'Data
Science', 'AI', 'SPSS'}
```

Removing Items from Sets

The discard( ) or remove( ) method can be used to remove a particular item from the set. The fundamental difference between discard( ) and remove( ) is that the first does not take any action if the item does not exist in the set, whereas remove( ) will raise an error in such scenario. Listing 1-32 gives example code for removing items from a set.

Listing 1-32. Example Code for Removing Items from a Set

```
# remove an element
languages.remove('AI')
print(languages)

# discard an element, although AI has already been removed discard will not
throw an error
languages.discard('AI')
print(languages)

# Pop will remove a random item from set
print ("Removed:", (languages.pop()), "from", languages)
---- output ----
{'Julia', 'Machine Learning', 'R', 'Python', 'SAS', 'Java', 'C', 'Data
Science', 'SPSS'}
{'Julia', 'Machine Learning', 'R', 'Python', 'SAS', 'Java', 'C', 'Data
Science', 'SPSS'}
Removed: Julia from {'Machine Learning', 'R', 'Python', 'SAS', 'Java', 'C',
'Data Science', 'SPSS'}
```

Set Operations

As discussed earlier, sets allow us to use mathematical set operations such as union, intersection, difference, and symmetric difference. We can achieve this with the help of operators or methods.

Set Unions

Union of two sets A and B will result in a set of all items combined from both sets. There are two ways to perform union operation: 1) using the | operator and 2) using the union( ) method. Refer to Listing 1-33 for a set union operation code example.

Listing 1-33. Example Code for Set Union Operation

```
# initialize A and B
A = {1, 2, 3, 4, 5}
B = {4, 5, 6, 7, 8}

# use | operator
print ("Union of A | B", A|B)

# alternative we can use union()
print ("Union of A | B", A.union(B))
---- output ----
Union of A | B {1, 2, 3, 4, 5, 6, 7, 8}
```

Set Intersections

The intersection of two sets A and B will result in a set of items that exist in common in both sets. There are two ways to achieve intersection operation: 1) using the and operator and 2) using the intersection( ) method. Refer to Listing 1-34 for set intersection operation example code.

Listing 1-34. Example Code for Set Intersection Operation

```
# use & operator
print ("Intersection of A & B", A & B)

# alternative we can use intersection()
print ("Intersection of A & B", A.intersection(B))
---- output ----
Intersection of A & B {4, 5}
```

Set Difference

The difference of two sets A and B (i.e., A - B) will result in a set of items that exists only in A and not in B. There are two ways to perform a difference operation: 1) using the '–,–' operator and 2) using the difference( ) method. Refer to Listing 1-35 for a set difference operation code example.

Listing 1-35. Example Code for Set Difference Operation

```
# use - operator on A
print ("Difference of A - B", A - B)

# alternative we can use difference()
print ("Difference of A - B", A.difference(B))
---- output ----
Difference of A - B {1, 2, 3}
```

Set Symmetric Difference

Symmetric difference of two sets A and B is a set of items from both sets that are not common. There are two ways to perform symmetric difference: 1) using a ^ operator and 2) using the symmetric_*difference( ) method*. Refer to Listing 1-36 for a set symmetric difference operation code example.

Listing 1-36. Example Code for Set Symmetric Difference Operation

```
# use ^ operator
print ("Symmetric difference of A ^ B", A ^ B)

# alternative we can use symmetric_difference()
print ("Symmetric difference of A ^ B", A.symmetric_difference(B))
---- output ----
Symmetric difference of A ^ B {1, 2, 3, 6, 7, 8}
```

Basic Operations

Let's look at fundamental operations that can be performed on Python sets in the Listing 1-37 code example.

Listing 1-37. Example Code for Basic Operations on Sets

```
# Return a shallow copy of a set
lang = languages.copy()
print (languages)
print (lang)
```

```
# initialize A and B
A = {1, 2, 3, 4, 5}
B = {4, 5, 6, 7, 8}

print (A.isdisjoint(B))     # True, when two sets have a null intersection
print (A.issubset(B))       # True, when another set contains this set
print (A.issuperset(B))     # True, when this set contains another set
sorted(B)                   # Return a new sorted list
print (sum(A))              # Retrun the sum of all items
print (len(A))              # Return the length
print (min(A))              # Return the largest item
print (max(A))              # Return the smallest item
---- output ----
{'Machine Learning', 'R', 'Python', 'SAS', 'Java', 'C', 'Data Science',
'SPSS'}
{'Machine Learning', 'R', 'Python', 'SAS', 'Java', 'C', 'Data Science',
'SPSS'}
False
False
False
15
5
1
5
```

Dictionary

Python dictionary will have a key and value pair for each item that is part of it. The key and value should be enclosed in curly braces. Each key and value are separated using a colon (:), and further, each item is separated by a comma (,). Note that the keys are unique within a specific dictionary and must be an immutable data type such as strings, numbers, or tuple, whereas values can take duplicate data of any type. Table 1-14 summarizes the Python dictionary operations; refer to Listings 1-38 through 1-42 for code examples.

Table 1-14. *Python Dictionary Operations*

Description	Python Expression	Example	Results
Creating a dictionary	dict = {'key1':'value1', 'key2':'value2'.....}	dict = {'Name': 'Jivin', 'Age': 8, 'Class': 'Three'}	{'Name': 'Jivin', 'Age': 8, 'Class': 'Three'}
Accessing items in dictionary	dict ['key']	dict['Name']	dict['Name']: Jivin
Deleting a dictionary	del dict['key']; dict.clear(); del dict;	del dict['Name']; dict.clear(); del dict;	{'Age':68, 'Class': 'Three'}; {};
Updating a dictionary	dict['key'] = new_value	dict['Age'] = 8.5	dict['Age']: 8.5
Length	len(dict)	len({'Name': 'Jivin', 'Age': 8, 'Class': 'Three'})	3
String representation of dict	str(dict)	dict = {'Name': 'Jivin', 'Age': 8}; print ("Equivalent String: ", str (dict))	Equivalent String : {'Age': 8, 'Name': 'Jivin'}
Returning the shallow copy of dict	dict.copy()	dict = {'Name': 'Jivin', 'Age': 8}; dict1 = dict.copy()print(dict1)	{'Age': 8, 'Name': 'Jivin'}
Creating a new dictionary with keys from seq and values set to value	dict.fromkeys()	seq = ('name', 'age', 'sex') dict = dict.fromkeys(seq) print ("New Dictionary: ", str(dict)) dict = dict.fromkeys(seq, 10) print ("New Dictionary: ", str(dict))	New Dictionary : {'age': None, 'name': None, 'sex': None} New Dictionary : {'age': 10, 'name': 10, 'sex': 10}
For key key, returns value or default if key not in dictionary	dict.get(key, default=None)	dict = {'Name': 'Jivin', 'Age': 8} print ("Value for Age: ", dict.get('Age'))print ("Value for Education: ", dict.get ('Education', "Third Grade"))	Value : 68 Value :Third Grade

(*continued*)

Table 1-14. (*continued*)

Description	Python Expression	Example	Results
Returns True if key in dictionary dict, False otherwise	dict.has_key(key)	dict = {'Name': 'Jivin', 'Age': 8} print ("Age exists? ", dict.has_key('Age')) print ("Sex exists? ", dict.has_key('Sex'))	Value : True Value : False
Returns a list of dict's (key, value) tuple pairs	dict.items()	dict = {'Name': 'Jivin', 'Age': 8} print ("dict items: ", dict.items())	Value : [('Age', 8), ('Name', 'Jivin')]
Returns list of dictionary dict's keys	dict.keys()	dict = {'Name': 'Jivin', 'Age': 8} print ("dict keys: ", dict.keys())	Value : ['Age', 'Name']
Similar to get(), but will set dict[key]=default if key is not already in dict	dict.setdefault(key, default=None)	dict = {'Name': 'Jivin', 'Age': 8} print ("Value for Age: ", dict.setdefault('Age', None)) print ("Value for Sex: ", dict.setdefault('Sex', None))	Value :8Value : None
Adds dictionary dict2's key-values pairs to dict	dict.update(dict2)	dict = {'Name': 'Jivin', 'Age': 8} dict2 = {'Sex': 'male' } dict.update(dict2) print("dict.update(dict2) =",dict)	Value : {'Age': 8, 'Name': 'Jivin', 'Sex': 'male'}
Returns list of dictionary dict's values	dict.values()	dict = {'Name': 'Jivin', 'Age': 8}print ("Value: ", dict.values())	Value : [8, 'Jivin']

Listing 1-38. Example Code for Creating a Dictionary

```
# Creating a dictionary
dict = {'Name': 'Jivin', 'Age': 8, 'Class': 'Three'}

print ("Sample dictionary: ", dict)
---- output ----
Sample dictionary:  {'Name': 'Jivin', 'Age': 8, 'Class': 'Three'}
```

Listing 1-39. Example Code for Accessing the Dictionary

```
print ("Value of key Name, from sample dictionary:", dict['Name'])
---- output ----
Value of key Name, from sample dictionary: Jivin
```

Listing 1-40. Example for Deleting a Dictionary

```
# Deleting a dictionary
dict = {'Name': 'Jivin', 'Age': 8, 'Class': 'Three'}
print ("Sample dictionary: ", dict)
del dict['Name'] # Delete specific item
print ("Sample dictionary post deletion of item Name:", dict)

dict = {'Name': 'Jivin', 'Age': 8, 'Class': 'Three'}
dict.clear() # Clear all the contents of dictionary
print ("dict post dict.clear():", dict)

dict = {'Name': 'Jivin', 'Age': 8, 'Class': 'Three'}
del dict # Delete the dictionary
---- output ----

Sample dictionary:  {'Name': 'Jivin', 'Age': 8, 'Class': 'Three'}
Sample dictionary post deletion of item Name: {'Age': 8, 'Class': 'Three'}
dict post dict.clear(): {}
```

Listing 1-41. Example Code for Updating the Dictionary

```
# Updating a dictionary

dict = {'Name': 'Jivin', 'Age': 8, 'Class': 'Three'}
print ("Sample dictionary: ", dict)
dict['Age'] = 8.5

print ("Dictionary post age value update: ", dict)
---- output ----
Sample dictionary:  {'Name': 'Jivin', 'Age': 8, 'Class': 'Three'}
Dictionary post age value update:  {'Name': 'Jivin', 'Age': 8.5, 'Class':
'Three'}
```

Listing 1-42. Example Code for Basic Operations on the Dictionary

```
# Basic operations

dict = {'Name': 'Jivin', 'Age': 8, 'Class': 'Three'}
print ("Length of dict: ", len(dict))

dict1 = {'Name': 'Jivin', 'Age': 8};
dict2 = {'Name': 'Pratham', 'Age': 9};
dict3 = {'Name': 'Pranuth', 'Age': 7};
dict4 = {'Name': 'Jivin', 'Age': 8};

# String representation of dictionary
dict = {'Name': 'Jivin', 'Age': 8}
print ("Equivalent String: ", str (dict))

# Copy the dict
dict1 = dict.copy()
print (dict1)

# Create new dictionary with keys from tuple and values to set value
seq = ('name', 'age', 'sex')

dict = dict.fromkeys(seq)
print ("New Dictionary: ", str(dict))
```

```python
dict = dict.fromkeys(seq, 10)
print ("New Dictionary: ", str(dict))

# Retrieve value for a given key
dict = {'Name': 'Jivin', 'Age': 8};
print ("Value for Age: ", dict.get('Age'))
# Since the key Education does not exist, the second argument will be
returned
print ("Value for Education: ", dict.get('Education', "First Grade"))

# Check if key in dictionary
print ("Age exists? ", 'Age' in dict)
print ("Sex exists? ", 'Sex' in dict)

# Return items of dictionary
print ("dict items: ", dict.items())

# Return items of keys
print ("dict keys: ", dict.keys())

# return values of dict
print ("Value of dict: ",  dict.values())

# if key does not exists, then the arguments will be added to dict and
returned
print ("Value for Age : ", dict.setdefault('Age', None))
print ("Value for Sex: ", dict.setdefault('Sex', None))

# Concatenate dicts
dict = {'Name': 'Jivin', 'Age': 8}
dict2 = {'Sex': 'male' }

dict.update(dict2)
print ("dict.update(dict2) = ",  dict)
---- output ----
Length of dict:  3
Equivalent String:  {'Name': 'Jivin', 'Age': 8}
{'Name': 'Jivin', 'Age': 8}
New Dictionary:  {'name': None, 'age': None, 'sex': None}
New Dictionary:  {'name': 10, 'age': 10, 'sex': 10}
```

```
Value for Age:  8
Value for Education:  First Grade
Age exists?  True
Sex exists?  False
dict items:  dict_items([('Name', 'Jivin'), ('Age', 8)])
dict keys:  dict_keys(['Name', 'Age'])
Value of dict:  dict_values(['Jivin', 8])
Value for Age :  8
Value for Sex:  None
dict.update(dict2) =  {'Name': 'Jivin', 'Age': 8, 'Sex': 'male'}
```

User-Defined Functions

A user-defined function is a block of related code statements that are organized to achieve a single related action. A key objective of the concept of the user-defined function is to encourage modularity and enable reusability of code.

Defining a Function

Functions need to be defined, and following are the set of rules to be followed to define a function in Python.

- The keyword def denotes the beginning of a function block that will be followed by the name of the function and open, close parentheses. After this, put a colon (:) to indicate the end of the function header.

- Functions can accept arguments or parameters. Any such input arguments or parameters should be placed within the parentheses in the header of the parameter.

- The main code statements are to be put below the function header and should be indented, which indicates that the code is part of the same function.

- Functions can return the expression to the caller. If the return method is not used at the end of the function, it will act as a subprocedure. The key difference between the function and the subprocedure is that the function will always return an expression, whereas the subprocedure will not.

Syntax for creating functions without argument:

```
def function_name():
    1st block line
    2nd block line
    ...
```

Refer to Listings 1-43 and 1-44 for user-defined function examples.

Listing 1-43. Example Code for Creating Functions Without Argument

```
# Simple function
def someFunction():
    print ("Hello World")

# Call the function
someFunction()
----- output -----
Hello world
```

Following is the syntax for creating functions with argument:

```
def function_name(parameters):
    1st block line
    2nd block line
    ...
    return [expression]
```

Listing 1-44. Example Code for Creating Functions with Arguments

```
# simple function to add two numbers
def sum_two_numbers(a, b):
    return a + b

# after this line x will hold the value 3!
x = sum_two_numbers(1,2)
print (x)

# You can also set default value for argument(s) in a function. In the
below example value of b is set to 10 as default
```

```python
def sum_two_numbers(a, b = 10):
    return a + b

print (sum_two_numbers(10))
print (sum_two_numbers(10, 5))
----- output -----
3
20
15
```

The Scope of Variables

The availability of a variable or identifier within the program during and after the execution is determined by the scope of a variable. There are two fundamental variable scopes in Python:

1. Global variables

2. Local variables

Refer to Listing 1-45 for code examples for defining variable scopes.

Note that Python does support global variables without you having to explicitly express that they are global variables.

Listing 1-45. Example Code for Defining Variable Scopes

```python
# Global variable
a = 10

# Simple function to add two numbers
def sum_two_numbers(b):
    return a + b

# Call the function and print result
print (sum_two_numbers(10))
----- output -----
20
```

Default Argument

You can define a default value for an argument of function, which means the function will assume or use the default value in case any value is not provided in the function call for that argument. Refer to Listing 1-46 for a code example.

Listing 1-46. Example Code for Function with Default Argument

```
# Simple function to add two number with b having default value of 10
def sum_two_numbers(a, b = 10):
    return a + b
# Call the function and print result
print (sum_two_numbers(10))
print (sum_two_numbers(10, 5))
----- output -----
20
15
```

Variable Length Arguments

There are situations when you do not know the exact number of arguments while defining the function, and would want the ability to process all the arguments dynamically. Python's answer for this situation is the variable length argument, which enables you to process more arguments than you specified while defining the function. The *args and **kwargs is a common idiom to allow a dynamic number of arguments.

The *args will provide all function parameters in the form of a tuple. Refer to Listings 1-47 and 1-48 for example code.

Listing 1-47. Example Code for Passing Arguments *args

```
# Simple function to loop through arguments and print them
def foo(*args):
    for a in args:
        print (a)

# Call the function
foo(1,2,3)
```

```
----- output -----
1
2
3
```

The **kwargs will give you the ability to handle named or keyword arguments keyword that you have not defined in advance.

Listing 1-48. Example Code for Passing Arguments as **kwargs

```
# Simple function to loop through arguments and print them
def foo(**kwargs):
    for a in kwargs:
        print (a, kwargs[a])

# Call the function
foo(name='Jivin', age=8)
----- output -----
name Jivin
age 8
```

Modules

A module is a logically organized, multiple, independent but related set of code or functions or classes. The key principles behind creating a module are that it's easier to understand and use, and has efficient maintainability. You can import a module and the Python interpreter will search for the module of interest in the following sequences.

First it searches the currently active directory, that is, the directory from which the Python your program is being called. If the module isn't found in the currently active directory, Python then searches each directory in the path variable PYTHONPATH. If this fails, it searches in the default package installation path

Note that the module search path is stored in the system module called sys as the sys.path variable, and this contains the current directory, PYTHONPATH, and the installation-dependent default.

When you import a module, it's loaded only once, regardless of the number of times it is imported. You can also import specific elements (functions, classes, etc.) from your module into the current namespace. Refer to Listing 1-49 for example code for importing modules.

Listing 1-49. Example Code for Importing Modules

```
# Import all functions from a module
import module_name         # Method 1
from modname import*        # Method 2

# Import specific function from the module
# Syntax: from module_name import function_name
from os import abc
```

Python internally has a dictionary known as the namespace that stores each variable or identifier name as the key, and their corresponding value is the respective Python object. There are two types of namespace, local and global. The local namespace gets created during the execution of a Python program to hold all the objects that are being created by the program. The local and a global variable have the same name, and the local variable shadows the global variable. Each class and function has its own local namespace. Python assumes that any variable assigned a value in a function is local. For global variables, you need to explicitly specify.

Another key built-in function is the dir( ); running this will return a sorted list of strings containing the names of all the modules, variables, and functions that are defined in a module. Refer to Listing 1-50 for example code.

Listing 1-50. Example Code dir( ) Operation

```
import os
content = dir(os)
print(content)

---- output ----
['DirEntry', 'F_OK', 'MutableMapping', 'O_APPEND', 'O_BINARY', 'O_CREAT',
'O_EXCL', 'O_NOINHERIT', 'O_RANDOM', 'O_RDONLY', 'O_RDWR', 'O_SEQUENTIAL',
'O_SHORT_LIVED', 'O_TEMPORARY', 'O_TEXT', 'O_TRUNC', 'O_WRONLY', 'P_
DETACH', 'P_NOWAIT', 'P_NOWAITO', 'P_OVERLAY', 'P_WAIT', 'PathLike',
'R_OK', 'SEEK_CUR', 'SEEK_END', 'SEEK_SET', 'TMP_MAX', 'W_OK', 'X_OK',
'_Environ', '__all__', '__builtins__', '__cached__', '__doc__', '__file__',
'__loader__', '__name__', '__package__', '__spec__', '_execvpe', '_exists',
'_exit', '_fspath', '_get_exports_list', '_putenv', '_unsetenv', '_wrap_
close', 'abc', 'abort', 'access', 'altsep', 'chdir', 'chmod', 'close',
```

```
'closerange', 'cpu_count', 'curdir', 'defpath', 'device_encoding',
'devnull', 'dup', 'dup2', 'environ', 'error', 'execl', 'execle', 'execlp',
'execlpe', 'execv', 'execve', 'execvp', 'execvpe', 'extsep', 'fdopen',
'fsdecode', 'fsencode', 'fspath', 'fstat', 'fsync', 'ftruncate', 'get_exec_
path', 'get_handle_inheritable', 'get_inheritable', 'get_terminal_size',
'getcwd', 'getcwdb', 'getenv', 'getlogin', 'getpid', 'getppid', 'isatty',
'kill', 'linesep', 'link', 'listdir', 'lseek', 'lstat', 'makedirs',
'mkdir', 'name', 'open', 'pardir', 'path', 'pathsep', 'pipe', 'popen',
'putenv', 'read', 'readlink', 'remove', 'removedirs', 'rename', 'renames',
'replace', 'rmdir', 'scandir', 'sep', 'set_handle_inheritable', 'set_
inheritable', 'spawnl', 'spawnle', 'spawnv', 'spawnve', 'st', 'startfile',
'stat', 'stat_result', 'statvfs_result', 'strerror', 'supports_bytes_
environ', 'supports_dir_fd', 'supports_effective_ids', 'supports_fd',
'supports_follow_symlinks', 'symlink', 'sys', 'system', 'terminal_size',
'times', 'times_result', 'truncate', 'umask', 'uname_result', 'unlink',
'urandom', 'utime', 'waitpid', 'walk', 'write']
```

Looking at the preceding output, __name__ is a special string variable name that denotes the module's name, and __file__ is the filename from which the module was loaded.

File Input/Output

Python provides easy functions to read and write information to a file (Table 1-15). To perform a read or write operation on a file, we need to open it first. Once the required operation is complete, it needs to be closed so that all the resources tied to that file are freed.

Following is the sequence of a file operation:

- Open a file

- Perform operations that are read or write

- Close the file

Table 1-15. *File Input/Output Operations*

Description	Syntax	Example
Opening a file	obj=open(filename , access_mode , buffer)	f = open('vehicles.txt', 'w')
Reading from a file	fileobject.read(value)	f = open('vehicles.txt') f.readlines()
Closing a file	fileobject.close()	f.close()
Writing to a file	fileobject.write(string str)	vehicles = ['scooter\n', 'bike\n', 'car\n'] f = open('vehicles.txt', 'w') f.writelines(vehicles) f.close()

Opening a File

While opening a file, the access_mode will determine the file open mode, that is, read, write, append, etc. Read (r) mode is the default file access mode and this is an optional parameter. Refer to Table 1-16 to learn about file opening modes and Listing 1-51 for example code.

Table 1-16. *File Opening Modes*

Modes	Description
R	Reading only
Rb	Reading only in binary format
r+	File will be available for both read and write
rb+	File will be available for both read and write in a binary format
W	Writing only
Wb	Writing only in binary format
w+	Open for both writing and reading; if file existing—overwrite, else—create
wb+	Open for both writing and reading in binary format; if file existing—overwrite, else— create
A	Opens file in append mode. Creates a file if it does not exist
Ab	Opens file in append mode. Creates a file if it does not exist
a+	Opens file for both append and reading. Creates a file if it does not exist
ab+	Opens file for both append and reading in binary format. Creates a file if it does not exist

Listing 1-51. Example Code for File Operations

```
# Below code will create a file named vehicles and add the items. \n is a
newline character
vehicles = ['scooter\n', 'bike\n', 'car\n']
f = open('vehicles.txt', 'w')
f.writelines(vehicles)
f.close

# Reading from file
f = open(vehicles.txt')
print (f.readlines())
f.close()

---- output ----

['scooter\n', 'bike\n', 'car\n']
```

Exception Handling

Any error that happens while a Python program is being executed that will interrupt the expected flow of the program is called an exception. Your program should be designed to handle both expected and unexpected errors.

Python has a rich set of built-in exceptions listed in Table 1-17 that force your program to output an error when something in it goes wrong.

Following is the list of Python Standard Exceptions as described in Python's official documentation (https://docs.python.org/2/library/exceptions.html)

Table 1-17. *Python Built-in Exception Handling*

Exception Name	Description
Exception	Base class for all exceptions
StopIteration	Raised when the next() method of an iterator does not point to any object
SystemExit	Raised by the sys.exit() function
StandardError	Base class for all built-in exceptions except StopIteration and SystemExit
ArithmeticError	Base class for all errors that occur for numeric calculation

(continued)

59

Table 1-17. (*continued*)

Exception Name	Description
OverflowError	Raised when a calculation exceeds the maximum limit for a numeric type
FloatingPointError	Raised when a floating point calculation fails
ZeroDivisonError	Raised when a division or modulo by zero takes place for all numeric types
AssertionError	Raised in case of failure of the Assert statement
AttributeError	Raised in case of failure of attribute reference or assignment
EOFError	Raised when there is no input from either the raw_input() or input() function and the end of file is reached
ImportError	Raised when an import statement fails
KeyboardInterrupt	Raised when the user interrupts program execution, usually by pressing Ctrl+c
LookupError	Base class for all lookup errors
IndexError	Raised when an index is not found in a sequence
KeyError	Raised when the specified key is not found in the dictionary
NameError	Raised when an identifier is not found in the local or global namespace
UnboundLocalError	Raised when trying to access a local variable in a function or method but no value has been assigned to it
EnvironmentError	Base class for all exceptions that occur outside the Python environment
IOError	Raised when an input/output operation fails, such as the print statement or the open() function when trying to open a file that does not exist
IOError	Raised for operating system-related errors
SyntaxError	Raised when there is an error in Python syntax
IndentationError	Raised when indentation is not specified properly
SystemError	Raised when the interpreter finds an internal problem, but when this error is encountered the Python interpreter does not exit
SystemExit	Raised when Python interpreter is quit by using the sys.exit() function. If not handled in the code, causes the interpreter to exit

(*continued*)

Table 1-17. (*continued*)

Exception Name	Description
TypeError	Raised when an operation or function is attempted that is invalid for the specified data type
ValueError	Raised when the built-in function for a data type has the valid type of arguments, but the arguments have invalid values specified
RuntimeError	Raised when a generated error does not fall into any category
NotImplementedError	Raised when an abstract method that needs to be implemented in an inherited class is not actually implemented

You can handle exceptions in your Python program using try, raise, except, and finally statements.

try and except: The try clause can be used to place any critical operation that can raise an exception in your program; the exception clause should have the code that will handle an exception. Refer to Listing 1-52 for example code for exception handling.

Listing 1-52. Example Code for Exception Handling

```
import sys

try:
    a = 1
    b = 1
    print ("Result of a/b: ", a / b)
except (ZeroDivisionError):
    print ("Can't divide by zero")
except (TypeError):
    print ("Wrong data type, division is allowed on numeric data type only")
```

```
except:
    print ("Unexpected error occurred", '\n', "Error Type: ", sys.exc_
    info()[0], '\n', "Error Msg: ", sys.exc_info()[1])
---- output ----
Result of a/b:   1.0
```

Note   1) Changing the value of b to zero in the preceding code will print the statement "Can't divide by zero."

2) Replacing "a" with "A" in the divide statement will print the following output:

Unexpected error occurred

Error Type: <type 'exceptions.NameError'>

Error Msg: name 'A' is not defined

Finally: This is an optional clause, which is intended to define clean-up actions that must be executed under all circumstances.

Refer to Listing 1-53 for example code for exception handling with file operations.

Listing 1-53. Example Code for Exception Handling with File Operations # Below code will open a file and try to convert the content to integer

```
try:
    f = open('C:\\Users\Manoh\\Documents\\ vehicles.txt')
    s = f.readline()
    print (s)
    i = int(s.strip())
except IOError as e:
    print ("I/O error({0}): {1}".format(e.errno, e.strerror))
except ValueError:
    print ("Could not convert data to an integer.")
except:
    print ("Unexpected error occurred", '\n', "Error Type: ", sys.exc_
info()[0], '\n', "Error Msg: ", sys.exc_info()[1])
```

```
finally:
    f.close()
    print ("file has been closed")
---- output ----
scooter
Could not convert data to an integer.
file has been closed
```

Python always executes a "finally" clause before leaving the try statement, irrespective of an exception occurrence. If an exception clause is not designed to handle the exception raised in the try clause, the same is reraised after the "finally" clause has been executed. Refer to Figure 1-2 for ideal code flow for an error handler. If usage of statements such as break, continue, or return forces the program to exit the try clause, the "finally" is still executed on the way out.

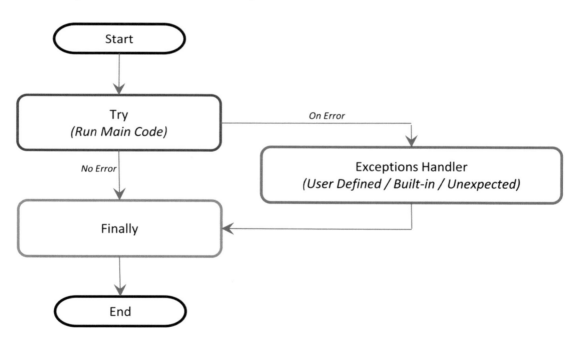

Figure 1-2. *Code flow for error handler*

Note that generally it's a best practice to follow the single exit point principle by using "finally." This means that either after successful execution of your main code or your error handler has finished handling an error, it should pass through the "finally" so that the code will be exited at the same point under all circumstances.

Summary

In this chapter I have tried to cover the basics and the essential topics to get you started in Python 3. There is an abundance of online/offline resource available to increase your knowledge depth about Python as a programming language. Table 1-18 provides some useful resources for your future reference.

Table 1-18. *Additional Resources*

Resource	Description	Mode
http://docs.python-guide. org/en/latest/intro/ learning/	This is Python's official tutorial; it covers all the basics and offers a detailed tour of the language and standard libraries.	Online
http://awesome-python.com/	A curated list of awesome Python frameworks, libraries, software and resources	Online
The Hacker's Guide to Python	This book is aimed at developers who already know Python but wants to learn from more experienced Python developers.	Book

CHAPTER 2

Step 2: Introduction to Machine Learning

Machine learning (ML) is a subfield of computer science that evolved from the study of *pattern recognition* and computational learning theory in artificial intelligence (AI). Let's look at a few other versions of definition that exist for ML:

- In 1959, Arthur Samuel, an American pioneer in the field of computer gaming, ML, and AI defined machine learning as a "Field of study that gives computers the ability to learn without being explicitly programmed."

- ML is a field of computer science that involves using statistical methods to create programs that either improve performance over time or detect patterns in massive amounts of data that humans would be unlikely to find.

The preceding definitions are correct. In short, ML is a collection of algorithms and techniques used to create computational systems that learn from data in order to make predictions and inferences.

ML applications are abounding. Let's look at some of the most common day-to-day applications of ML that happen around us.

Recommendation system: YouTube suggests videos for each of its users that it believes the individual user will be interested in, based on a recommendation system. Similarly, Amazon and other such e-retailers suggest a product in which a customer will be interested and likely to purchase, by looking at the purchase history for the customer and a large inventory of products.

Spam detection: E-mail service providers use an ML model that can automatically detect and move the unsolicited messages to the spam folder.

© Manohar Swamynathan 2019
M. Swamynathan, *Mastering Machine Learning with Python in Six Steps*,
https://doi.org/10.1007/978-1-4842-4947-5_2

Prospect customer identification: Banks, insurance companies, and financial organizations use ML models that trigger alerts so those organizations intervene at the right time to start engaging customers with the right offers and persuade them to convert early. These models observe the pattern of behavior by a user during the initial period and map it to the past behaviors of all users, attempting to identify those who will buy the product and those who will not.

In this chapter we will learn about the history and evolution of ML to understand where it fits in the wider AI family. We'll also learn about different related forms/terms such as statistics, data or business analytics, and data science that exist parallel to ML and why they exist. Also discussed are high level categories of ML, and the most commonly used frameworks to build efficient ML systems. We'll also briefly look at the key ML libraries for data analysis.

History and Evolution

ML is a subset of AI, so let's first understand what AI is and where ML fits within its wider umbrella. AI is a broad term that aims to use data to offer solutions to existing problems. It is the science and engineering of replicating, even surpassing, human-level intelligence in machines. That means to observe or read, learn, sense, and experience.

The AI process loop is shown in Figure 2-1.

Figure 2-1. *AI process loop*

- *Observe*: Identify patterns using the data.

- *Plan*: Find all possible solutions.

- *Optimize*: Find optimal solution form the list of possible solutions.

- *Action*: Execute the optimal solution.

- *Learn and Adapt*: Is the result giving the expected result? if not, adapt.

The AI process loop can be achieved using intelligent agents. A robotic intelligent agent can be defined as a component that can perceive its environment through different kinds of sensors (camera, infrared, etc.), and will take actions within the environment. Here, robotic agents are designed to reflect humans. We have different sensory organs such as eyes, ears, nose, tongue, and skin to perceive our environment, and organs such as hands, legs, and mouth are the effectors that enable us to take action within our environment based on our perceptions

A detailed discussion on designing the agent is discussed in the book *Artificial Intelligence, A Modern Approach* by Stuart J. Russell and Peter Norvig. Figure 2-2 is a sample pictorial representation.

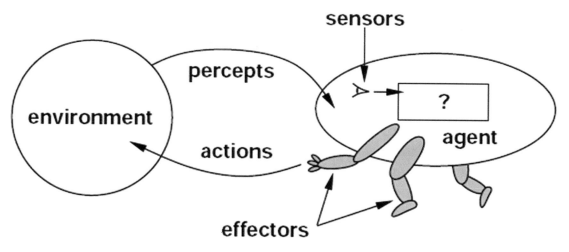

Figure 2-2. *Depiction of a robotic intelligent agent concept that interacts with its environment through sensors and effectors*

To get a better understanding of the concept, let's look at the components of an intelligent agent designed for a particular environment or use case (Table 2-1). Consider designing an automated taxi driver.

Table 2-1. *Example of Intelligent Agent Components*

Intelligent Agent's Component Name	Description
Agent Type	Taxi driver
Goals	Safe trip, legal, comfortable trip, maximize profits, convenient, fast
Environment	Roads, traffic, signals, signage, pedestrians, customers
Percepts	Speedometer, microphone, GPS, cameras, sonar, sensors
Actions	Steer, accelerate, brake, talk to a passenger

The taxi driver robotic intelligent agent will need to know its location, the direction in which it's traveling, the speed at which it's traveling, and what else is on the road! This information can be obtained from the percepts such as controllable cameras in appropriate places, the speedometer, odometer, and accelerometer. For understanding the mechanical state of the vehicle and engine, it needs electrical system sensors. In addition, a satellite global positioning system (GPS) can help to provide accurate position information with respect to an electronic map, and infrared/sonar sensors can detect distances to other cars or obstacles around it. The actions available to the intelligent taxi driver agent are control over the engine through the pedals for accelerating and braking, and steering for controlling direction. There should also be a way to interact with or talk to the passengers, to understand the destination or goal.

In 1950, Alan Turing a well-known computer scientist, proposed a test known as the *Turing test* in his famous paper "Computing Machinery and Intelligence." The test was designed to provide a satisfactory operational definition of intelligence, which required that a human being should not be able to distinguish the machine from another human being by using the replies to questions put to both.

To be able to pass the Turing test, the computer should possess the following capabilities:

- *Natural language processing*: To be able to communicate successfully in a chosen language

- *Knowledge representation*: To store information provided before or during the interrogation that can help in finding information, making decisions, and planning. This is also known as an Expert System.

- *Automated reasoning* (speech): To use the stored knowledge map information to answer questions and to draw new conclusions where required

- *Machine learning*: Analyzing data to detect and extrapolate patterns that will help adapt to new circumstances

- *Computer vision*: To perceive objects or the analyzing of images to find features of the images

- *Robotics*: Devices that can manipulate and interact with their environment. That means to move objects around based on the circumstances.

- *Planning, scheduling, and optimization*: Figuring the ways to make decision plans or achieve specified goals, as well as analyzing the performance of the plans and designs

The aforementioned seven capability areas of AI have seen a great deal of research and growth over the years. Although many of the terms in these areas are used interchangeably, we can see from the description that their objectives are different (Figure 2-3). Particularly, ML has the scope to cut across all seven areas of AI.

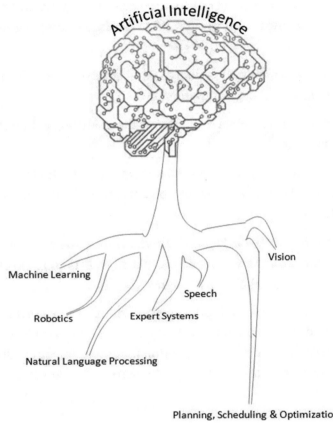

Figure 2-3. *AI areas*

Artificial Intelligence Evolution

Let's briefly look at the past, present, and future of AI.

Past

Current

Future

Artificial Narrow Intelligence (ANI): Machine intelligence that equals or exceeds human intelligence or efficiency at a specific task. An example is IBM's Watson, which requires close participation of subject matter or domain experts to supply data/information and evaluate its performance.

Artificial General Intelligence (AGI): A machine with the ability to apply intelligence to any problem for an area, rather than just one specific problem. Self-driving cars are a good example of this.

Artificial Super Intelligence (ASI): An intellect that is much smarter than the best human brains in practically every field, general wisdom, social skills, and including scientific creativity. The key theme here is "don't model the world, model the mind."

Different Forms

Is ML the only subject in which we use data to learn and use for prediction/inference?

To answer that question, let's have a look at the definition (Wikipedia) of the few other key terms (not an exhaustive list) that are often heard relatively:

- *Statistics:* It is the study of the collection, analysis, interpretation, presentation, and organization of data.

- *Data mining:* It is an interdisciplinary subfield of computer science. It is the computational process of discovering patterns in large data sets (from a data warehouse) involving methods at the intersection of AI, ML, statistics, and database systems.

- *Data analytics:* It is a process of inspecting, cleaning, transforming, and modeling data, with the goal of discovering useful information, suggesting conclusions, and supporting decision making. This is also known as business analytics and is widely used in many industries to allow companies/organization to use the science of examining raw data for the purpose of drawing conclusions about that information to make better business decisions.

- *Data science:* Data science is an interdisciplinary field about processes and systems to extract knowledge or insights from data in various forms, either structured or unstructured, which is a continuation of some of the data analysis fields such as statistics, ML, data mining, and predictive analytics, similar to knowledge discovery in databases (KDD).

Yes, from the preceding definitions it is clear and surprising to find out that ML isn't the only subject in which we use data to learn from it and use further for prediction/inference. The almost identical theme, tools, and techniques are being talked about in each of these areas. This raises a genuine question about why there are so many different names with a lot of overlap around learning from data. What is the difference between these?

The short answer is that all of these are practically the same. However, there exists a subtle difference or shade of meaning, expression, or sound between each of these. To get a better understanding, we'll have to go back to the history of each of these areas and closely examine the origin, core area of application, and evolution of these terms.

Statistics

German Scholar Gottfried Achenwall introduced the word "statistics" in the middle of the 18th century (1749). Usage of this word during this period meant that it was related to the administrative functioning of a state, supplying the numbers that reflect the periodic actuality regarding its various areas of administration. The origin of the word statistics may be traced to the Latin word "Status" ("council of state") or the Italian word "Statista" ("statesman" or "politician"), that is, the meaning of these words is "political State" or a Government. Shakespeare used a word statist in his drama Hamlet (1602). In the past, statistics were used by rulers, which designated the analysis of data about the state, signifying the "science of state."

At the beginning of the 19th century, statistics attained the meaning of the collection and classification of data. The Scottish politician Sir John Sinclair introduced it to English in 1791 in his book *Statistical Account of Scotland.* Therefore the fundamental purpose of the birth of statistics concerned data to be used by government and centralized administrative organizations to collect censuses data about the population for states and localities.

Frequentist

John Graunt was one of the first demographers and our first vital statistician. He published his observations in the *Bills of Mortality* (in 1662), and this work is often quoted as the first instance of descriptive statistics. He presented a vast amount of data in a few tables, which can be easily comprehended, and this technique is now widely known as descriptive statistics. In it, we note that weekly mortality statistics first appeared in England in 1603 at the Parish Clerks Hall. We can learn from it that in 1623, of some 50,000 burials in London, only 28 died of the plague. By 1632, this disease had practically disappeared, for the time being, to reappear in 1636 and again in the terrible epidemic of 1665. This exemplifies that the fundamental nature of descriptive statistics is counting. From all the parish registers, he counted the number of persons who died, and who died of the plague. The counted numbers every so often were relatively too large to follow, so he also simplified them by using proportion rather than the actual number. For example, the year 1625 had 51,758 deaths, of which 35,417 were of the plague. To simplify this he wrote, "We find the plague to bear unto the whole in proportion as 35 to 51. Or 7 to 10." With these, he is introducing the concept that the relative proportions are often of more interest than the raw numbers. We would generally express the proportion as 70%. This type of conjecture that is based on a sample data's proportion spread or frequency is known as "frequentist statistics." Statistical hypothesis testing is based on an inference framework, where you assume that observed phenomena are caused by the unknown but fixed process.

Bayesian

In contrast, Bayesian statistics (named after Thomas Bayes), describes the probability of an event, based on conditions that might be related to the event. At the core of Bayesian statistics is Bayes' theorem, which describes the outcome probabilities of related (dependent) events using the concept of conditional probability. For example, if a particular illness is related to age and lifestyle, then by applying Bayes' theorem by considering a person's age and lifestyle the probability of that individual having the illness can be assessed more accurately.

Bayes theorem is stated mathematically as the following equation:

$$P(A|B) = \frac{P(B|A)\ P(A)}{P(B)}$$

where A and B are events and P (B) $\neq$ 0.

- P (A) and P (B) are the probabilities of observing A and B without regard to each other.

- P (A | B), a conditional probability, is the probability of observing event A given that B is true.

- P (B | A) is the probability of observing event B given that A is true.

For example, a doctor knows that lack of sleep causes a migraine 50% of the time. The prior probability of any patient having lack of sleep is 10,000/50,000 and the prior probability of any patient having a migraine is 300/1,000. If a patient has a sleep disorder, let's apply Bayes theorem to calculate the probability of he/she having a migraine.

P (Sleep disorder | Migraine) = P(Migraine | Sleep disorder) * P(Migraine) / P(Sleep disorder)
P (Sleep disorder | Migraine) = .5 * 10000/50000 / (300/1000) = 33%

In the preceding scenario, there is a 33% chance that a patient with a sleep disorder will have a migraine problem.

Regression

Another major milestone for statisticians was the regression method, which was published by Legendre in 1805 and by Gauss in 1809. Legendre and Gauss both applied the method to the problem of determining, from astronomical observations, the orbits of bodies about the Sun—mostly comets but also, later, the then newly discovered minor planets. Gauss published a further development of the theory of least squares in 1821. Regression analysis is an essential statistical process for estimating the relationships between factors. It includes many techniques for analyzing and modeling various factors. The main focus here is about the relationship between a dependent factor and one or many independent factors, also called predictors or variables or features. We'll learn about this more in the fundamentals of ML with Scikit-learn.

Over time, the idea behind the word statistics has undergone an extraordinary transformation. The character of data or information provided has been extended to all spheres of human activity. Let's understand the difference between two terms quite often used along with statistics: 1) data and 2) method. Statistical data is the numerical statement of facts, whereas statistical methods deal with information on the principles and techniques used in collecting and analyzing such data. Today, statistics as a separate discipline from mathematics is closely associated with almost all branches of education and human endeavor that are mostly numerically representable. In modern times, it has innumerable and varied applications both qualitatively and quantitatively. Individuals and organizations use statistics to understand data and make informed decisions throughout the natural and social sciences, medicine, business, and other areas. Statistics has served as the backbone and given rise to many other disciplines, which you'll understand as you read further.

Data Mining

The term "knowledge discovery in databases" (KDD) was coined by Gregory Piatetsky-Shapiro in 1989. At the same time, he cofounded the first workshop named KDD. The term "data mining" was introduced in the 1990s in the database community, but data mining is the evolution of a field with a slightly longer history.

Data mining techniques are the result of research on the business process and product development. This evolution began when business data was first stored on computers in the relational databases and continued with improvements in data access, and further produced new technologies that allow users to navigate through their data in real time. In the business community, data mining focuses on providing "right data" at the "right time" for "right decisions". This is achieved by enabling tremendous data collection and applying algorithms on them with the help of distributed multiprocessor computers, to provide real-time insights from data.

We'll learn more about the five stages proposed by KDD for data mining in the "Framework for Building ML Systems" section.

Data Analytics

Analytics have been known to be used in business since the time of management movements toward industrial efficiency that were initiated in the late 19th century by Frederick Winslow Taylor, an American mechanical engineer. The manufacturing industry adopted measuring the pacing of the manufacturing and assembly line, as a result revolutionizing industrial efficiency. But analytics began to command more awareness in the late 1960s when computers had started playing a dominating role in organizations' decision support systems. Traditionally, business managers were making decisions based on past experiences or rules of thumb, or there were other qualitative aspects to decision making. However, this changed with the development of data warehouses and enterprise resource planning (ERP) systems. The business managers considered data and relied on ad hoc analysis to affirm their experience/knowledge-based assumptions for daily and critical business decisions. This evolved as data-driven business intelligence or business analytics for the decision-making process and was fast adopted by organizations and companies across the globe. Today, businesses of all sizes use analytics. Often the term "business analytics" is used interchangeably for "data analytics" in the corporate world.

Businesses need to have a holistic view of the market and how a company competes efficiently within that market to increase their RoI (return on investment). This requires a robust analytic environment around the kinds of analytics that are possible. These can be broadly categorized into four types (Figure 2-4).

1. Descriptive analytics

2. Diagnostic analytics

3. Predictive analytics

4. Prescriptive analytics

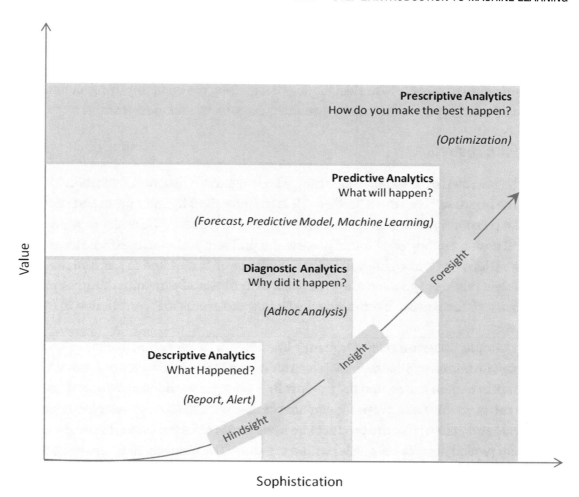

Figure 2-4. *Data analytics types*

Descriptive Analytics

They are the analytics that describe the past to tell us "what has happened." To elaborate, as the name suggests any activity or method that helps us to describe or summarize raw data into something interpretable by humans can be termed as descriptive analytics. These are useful because they allow us to learn from past behaviors, and understand how they might influence future outcomes.

Statistics such as the arithmetic operation of the count, min, max, sum, average, percentage, percent change, etc. fall into this category. Common examples of descriptive analytics are a company's business intelligence reports that cover different aspects of the organization to provide historical hindsight regarding the company's production, operations, sales, revenue, financials, inventory, customers, and market share.

Diagnostic Analytics

This is the next step to descriptive analytics, which examines data or information to answer the question "Why did it happen?" It is characterized by techniques such as drill down, data discovery, data mining, correlations, and causation. It basically provides a very good understanding of a limited piece of the problem you want to solve. However, it is very laborious work, as significant human intervention is required to perform the drill down or data mining to go deeper into the data to understand why something happened or root cause. It focuses on determining the factors and events that contributed to the outcome.

For example, assume a retail company's hardlines (a category that usually encompass furniture, appliance, tools, electronics, etc.) sales performance is not up to the mark in certain stores, and the product line manager would like to understand the root cause. In this case, the product manager may want to look backward to review past trends and patterns for the product line sales across different stores, based on its placement (which floor, corner, aisle) within the store. The manager may also want to understand if there is any causal relationship with other products that are closely kept with it. They may look at different external factors such as demographics, season, and macroeconomic factors separately as well as in unison to define relative ranking of the related variables based on concluded explanations. To accomplish this there is not a clearly defined set of ordered steps defined, and it depends on the experience level and thinking style of the person carrying out the analysis.

There is significant involvement of the subject matter expert, and the data/information may need to be presented visually for better understanding. There is a plethora of tools available, such as Excel, Tableau, QlikView, Spotfire, and D3 that enable diagnostic analytics.

Predictive Analytics

It is the ability to make predictions or estimation of likelihood about unknown future events based on the past or historical patterns. Predictive analytics will give us insight into "What might happen?" It uses many techniques from data mining, statistics, modeling, ML, and AI to analyze current data to make predictions about the future.

It is important to remember that the foundation of predictive analytics is based on probabilities, and the quality of prediction by statistical algorithms depends a lot on the quality of input data. Hence these algorithms cannot predict the future with 100% certainty. However, companies can use these statistics to forecast the probability of what might happen in the future, and considering these results alongside business knowledge should result in profitable decisions.

ML is heavily focused on predictive analytics, where we combine historical data from different sources such as organizational ERP, CRM (customer relationship management), POS (point of sale), employee data, and market research data. This data is used to identify patterns and apply statistical model/algorithms to capture relationships between various datasets and further predict the likelihood of an event.

Some examples of predictive analytics are weather forecasting, e-mail spam identification, fraud detection, the probability of a customer purchasing a product or renewal of insurance policy, predicting the chances of a person with known illness, etc.

Prescriptive Analytics

It is the area of data or business analytics dedicated to finding the best course of action for a given situation. Prescriptive analytics is related to the other three forms of analytics: descriptive, diagnostic, and predictive. The endeavor of prescriptive analytics is to measure a future decision's effect to enable the decision makers to foresee the possible outcomes before the actual decisions are made. Prescriptive analytic systems are a combination of business rules and ML algorithms, tools that can be applied against historic and real-time data feed. The key objective here does not just predict what will happen but also why it will happen, by predicting multiple futures based on different scenarios to allow companies to asses possible outcomes based on their actions.

An example of prescriptive analytics is using simulation in design situations to help users identify system behaviors under different configurations. This ensures all key performance metrics are met such as wait times, queue length, etc. Another example is to use linear or nonlinear programming to identify the best outcome for the business, given constraints and objective function.

Data Science

In 1960, Peter Naur used the term "data science" in his publication *Concise Survey of Computer Methods*, which is about contemporary data processing methods in a wide range of applications. In 1991, computer scientist Tim Berners-Lee announced the birth of what would become the World Wide Web as we know it today, in a post in the "Usenet group" he sets out the specifications for a worldwide, interconnected web of data, accessible to anyone from anywhere. Over time the Web/Internet has been growing tenfold each year and has become a global computer network providing a variety of information and communication facilities, consisting of interconnected networks using standardized communication protocols. Alongside, the storage systems also evolved and digital storage became more cost effective than paper.

As of 2008, the world's servers processed 9.57 zeta-bytes (9.57 trillion gigabytes) of information, which is equivalent to 12 gigabytes of information per person per day, according to the "How Much Information? 2010 report on Enterprise Server Information."

The rise of the Internet drastically increased the volume of structured, semistructured, and unstructured data. This led to the birth of the term "big data," characterized by three Vs (Figure 2-5): volume, variety, and velocity. Special tools and systems are required to process a high volume of data, with a wide variety (text, number, audio, video, etc), generated at a high velocity.

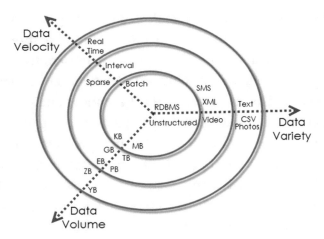

Figure 2-5. *Three Vs of big data (source: http://blog.sqlauthority.com)*

The big data revolution influenced the birth of the term "data science." Although the term data science existed from 1960, it became popular and is attributed to Jeff Hammerbacher and DJ Patil of Facebook and LinkedIn because they carefully chose it, attempting to describe their teams and work (as per *Building Data Science Teams* by DJ Patil, published in 2008). They settled on "data scientist" and a buzzword was born. Figure 2-6 is one picture that explains well the essential skills set for data science that was presented by Drew Conway in 2010.

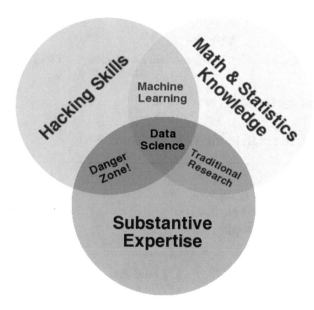

Figure 2-6. *Drew Conway's data science Venn diagram*

Executing data science projects requires three key skills:

- Programming or hacking skills

- Math and statistics

- Business or subject matter expertise for a given area in scope

---

Note that ML originated from AI. It is not a branch of data science, which rather only uses ML as a tool.

Statistics vs. Data Mining vs. Data Analytics vs. Data Science

We can learn from the history and evolution of subjects concerning learning from data that even though they use the same methods, they evolved as different cultures, so they have different histories, nomenclature, notation, and philosophical perspectives (Figure 2-7).

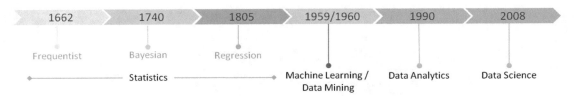

Figure 2-7. *The learn from data evolution*

All forms together: the path to ultimate AI (Figure 2-8).

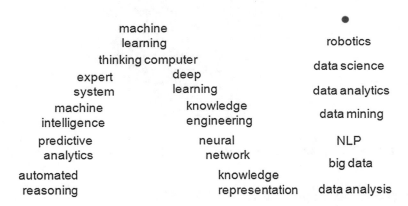

Figure 2-8. *All forms together: the path to ultimate AI*

Machine Learning Categories

At a high level, ML tasks can be categorized into three groups (Figure 2-9), based on the desired output and the kind of input required to produce it.

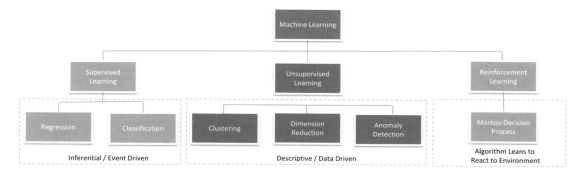

Figure 2-9. *Types of ML*

Supervised Learning

The ML algorithm is provided with a large enough example input dataset respective of output or event/class, usually prepared in consultation with the subject matter expert of a respective domain. The goal of the algorithm is to learn patterns in the data and build a general set of rules to map input to the class or event.

Broadly there are two types of commonly used supervised learning algorithms:

- *Regression*: The output to be predicted is a continuous number in relevance with a given input dataset. Example use cases are a prediction of retail sales, prediction of the number of staff required for each shift, number of car park spaces required for a retail store, a credit score for a customer, etc.

- *Classification*: The output to be predicted is the actual or the probability of an event/class and the number of classes to be predicted can be two or more. The algorithm should learn the patterns in the relevant input of each class from historical data and be able to predict the unseen class or event in the future, considering their input. An example use case is spam e-mail filtering where the output expected is to classify an e-mail into either spam or not spam.

Building supervised learning ML models has three stages:

1. *Training*: The algorithm will be provided with historical input data with the mapped output. The algorithm will learn the patterns within the input data for each output and represent that as a statistical equation, which is also commonly known as a model.

2. *Testing or validation*: In this phase, the performance of the trained model is evaluated, usually by applying it on a dataset (that was not used as part of the training) to predict the class or event.

3. *Prediction*: Here we apply the trained model to a data set that was not part of either the training or testing. The prediction will be used to drive business decisions.

Unsupervised Learning

There are situations where the desired output class/event is unknown for historical data. The objective of such cases would be to study the patterns in the input dataset to get a better understanding and identify similar patterns that can be grouped into specific classes or events. As these types of algorithms do not require any intervention from the subject matter experts beforehand, they are called unsupervised learning.

Following are some examples of unsupervised learning:

- *Clustering*: Assume that the classes are not known beforehand for a given dataset. The goal here is to divide the input dataset into logical groups of related items. Some examples are grouping similar news articles or grouping similar customers based on their profile.

- *Dimension reduction*: Here the goal is to simplify a large input dataset by mapping them to a lower dimensional space. For example, carrying analysis on a large dimension data set is very computationally intensive; so to simplify, you may want to find the key variables that hold a significant percentage (say 95%) of information and only use *them* for analysis.

Reinforcement Learning

The basic objective of reinforcement learning algorithms is to map situations to actions that yield the maximum final reward. While mapping the action, the algorithm should not just consider the immediate reward but also the next and all subsequent rewards. For example, a program to play a game or drive a car will have to constantly interact with a dynamic environment in which it is expected to achieve a certain goal. We'll learn more about this in detail later in Chapter 6.

Examples of reinforcement learning techniques are:

- Markov decision process

- Q-learning

- Temporal difference methods

- Monte-Carlo methods

Frameworks for Building ML Systems

Over time, the data mining field has seen a massive expansion. There have been a lot of efforts made by many experts to standardize methodologies and define best practice for the ever-growing, diversified, and iterative process of building ML systems. Over the last decade, the field of ML has become very important for different industries, businesses, and organizations because of its ability to extract insight from a huge amount of data. This data previously had no use or was underutilized to learn the trends/patterns and predict the possibilities that help to drive business decisions leading to profit. Ultimately, the risk of wasting the valuable information contained by the rich business data sources was raised. This required the use of adequate techniques to get useful knowledge, and the field of ML had emerged in the early 1980s and has seen great growth. With the emergence of this field, different process frameworks were introduced. These process frameworks guide and carry the ML tasks and their applications. Efforts were made to use data mining process frameworks that will guide the implementation of data mining on a huge amount of data.

Mainly three data mining process frameworks have been the most popular and widely practiced by data mining experts/researchers to build ML systems. These models are:

- Knowledge discovery in databases (KDD) process model

- Cross-industry standard process for data mining (CRISP-DM)

- Sample, Explore, Modify, Model, and Assess (SEMMA)

Knowledge Discovery in Databases

It refers to the overall process of discovering useful knowledge from data, which was presented by Fayyad et al. in 1996. It is an integration of multiple technologies for data management such as data warehousing, statistic ML, decision support, visualization, and parallel computing. As the name suggests, KDD centers on the overall process of knowledge discovery from data which covers the entire life cycle of data. That includes how the data is stored, how it is accessed, how algorithms can be scaled to enormous datasets efficiently, and how results can be interpreted and visualized.

Figure 2-10 shows the five stages in KDD, which are detailed in the following sections.

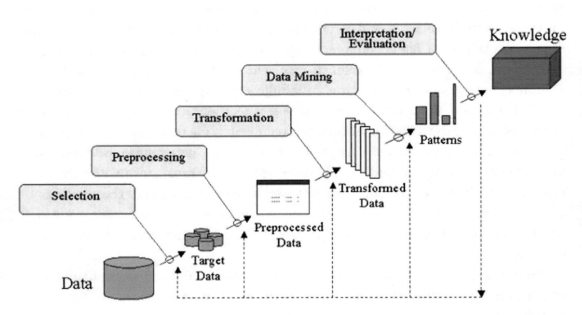

Figure 2-10. *KDD data mining process flow*

Selection

In this step, the selection and integration of the target data from possibly many different and heterogeneous sources are performed. Then the correct subset of variables and data samples *relevant to the analysis task is retrieved from the database.*

Preprocessing

Real-world datasets are often incomplete. That is, attribute values will be missing, noisy (errors and outliers), and inconsistent, which means there exist discrepancies between the collected data. The unclean data can confuse the mining procedures and lead to unreliable and invalid outputs. Also, performing complex analysis and mining on a huge amount of such soiled data may take a very long time. Preprocessing and cleaning should improve the quality of data and mining results by enhancing the actual mining process. The actions to be taken include the following:

- Collecting required data or information to model

- Outlier treatment or removal of noise

- Using prior domain knowledge to remove the inconsistencies and duplicates from the data

- Choosing strategies for handling missing data

Transformation

In this step, data is transformed or consolidated into forms appropriate for mining, that is, finding useful features to represent the data depending on the goal of the task. For example, in high-dimensional spaces or a large number of attributes, the distances between objects may become meaningless. So dimensionality reduction and transformation methods can be used to reduce the effective number of variables under consideration or find invariant representations for the data. There are various data transformation techniques:

- Smoothing (binning, clustering, regression, etc.

- Aggregation

- Generalization, in which a primitive data object can be replaced by higher level concepts

- Normalization, which involves min-max-scaling or z-score

- Feature construction from the existing attributes Principal Component Analysis (PCA), Multi Dimensional Scaling (MDS)

- Data reduction techniques are applied to produce a reduced representation of the data (the smaller volume that closely maintains the integrity of the original data)

- Compression, e.g., wavelets, PCA, clustering

Data Mining

In this step, ML algorithms are applied to extract data patterns. Exploration/ summarization methods such as mean, median, mode, standard deviation, a class/ concept description, and graphical techniques of low-dimensional plots can be used to understand the data. Predictive models such as classification or regression can be used to predict the event or future value. Cluster analysis can be used to understand the existence of similar groups. Select the most appropriate methods to be used for the model and pattern search.

Interpretation / Evaluation

This step is focused on interpreting themed patterns to make them understandable by the user, such as summarization and visualization. The mined pattern or models are interpreted. Patterns are a local structure that makes statements only about restricted regions of the space spanned by the variables. Models are global structures that make statements about any point in measurement space, e.g., $Y = mX+C$ (linear model).

Cross-Industry Standard Process for Data Mining

It is generally known by its acronym, CRISP-DM. It was established by the European Strategic Program on Research in Information Technology initiative, with an aim to create an unbiased methodology that is not domain dependent. It is an effort to consolidate data mining process best practices followed by experts to tackle data mining problems. It was conceived in 1996 and first published in 1999, and was reported as the leading methodology for data mining/predictive analytics projects in polls conducted in 2002, 2004, and 2007. There was a plan between 2006 and 2008 to update CRISP-DM but that update did not take place, and today the original CRISP-DM.org website is no longer active.

This framework is an idealized sequence of activities. It is an iterative process, and many of the tasks backtrack to previous tasks and repeat certain actions to bring more clarity. There are six major phases, as shown in Figure 2-11 and discussed in the following sections:

- Business understanding
- Data understanding

- Data preparation

- Modeling

- Evaluation

- Deployment

Figure 2-11. *Process diagram showing the relationship between the six phases of CRISP-DM*

Phase 1: Business Understanding

As the name suggests, the focus at this stage is to understand the overall project objectives and expectations from a business perspective. These objectives are converted to a data mining or ML problem definition and a plan of action is designed around data requirement, business owners input, and outcome performance evaluation metrics.

Phase 2: Data Understanding

In this phase, initial data is collected that was identified as a requirement in the previous phase. Activities are carried out to understand data gaps or relevance of the data to the object in hand, any data quality issues, and first insights into the data to bring out appropriate hypotheses. The outcome of this phase will be presented to the business iteratively, to bring more clarity into the business understanding and project objective.

Phase 3: Data Preparation

This phase is all about cleaning the data so that it's ready to be used for the model building phase. Cleaning data could involve filling the known data gaps from the previous step, missing value treatment, identifying the important features, applying transformations, and creating new relevant features where applicable. This is one of the most important phases, as the model's accuracy will depend significantly on the quality of data that is being fed into the algorithm to learn the patterns.

Phase 4: Modeling

There are multiple ML algorithms available to solve a given problem. So various appropriate ML algorithms are applied to the clean dataset and their parameters are tuned to the optimal possible values. Model performance for each of the applied models is recorded.

Phase 5: Evaluation

In this stage, a benchmarking exercise will be carried out among all the different models that were identified as giving high accuracy. The model will be tested against data that was not used as part of the training, to evaluate its performance consistency. The results will be verified against the business requirement identified in phase 1. The subject matter experts from the business will be involved to ensure that the model results are accurate and usable as required by the project objective.

Phase 6: Deployment

The key focus in this phase is the usability of the model output. So the final model signed off by the subject matter experts will be implemented, and the consumers of the model output will be trained on how to interpret or use it to make the business decisions defined in the business understanding phase. The implementation could be generating a prediction report and sharing it with the consumers. Also, periodic model training and prediction times will be scheduled, based on the business requirement.

SEMMA (Sample, Explore, Modify, Model, Assess)

SEMMA is the sequential steps to build ML models incorporated in SAS Enterprise Miner, a product by SAS Institute Inc., one of the largest producers of commercial, statistical, and business intelligence software. The sequential steps guide the development of an ML system. Let's look at the five sequential steps to understand better.

Sample

This step is all about selecting a subset of the right volume from a large dataset provided for building the model. This will help to build the model efficiently. This was a famous practice when computation power was expensive; however, it's still in practice. The selected subset of data should be an actual representation of the entire dataset originally collected, which means it should contain sufficient information to retrieve. The data is also divided for training and validation at this stage.

Explore

In this phase, activities are carried out to understand the data gaps and relationships among variables. Two key activities are univariate and multivariate analysis. In univariate analysis, each variable is examined individually to understand its distribution, whereas in multivariate analysis the relationship between each variable is explored. Data visualization is heavily used to help understand the data better.

Modify

In this phase, variables are cleaned where required. Newly derived features are created by applying business logic to existing features based on the requirement. Variables are transformed if necessary. The outcome of this phase is a clean dataset that can be passed to the ML algorithm to build the model.

Model

In this phase, various modeling or data mining techniques are applied to the preprocessed data, to benchmark their performance against the desired outcome.

Assess

This is the last phase. Here model performance is evaluated against the test data (not used in model training) to ensure reliability and business usefulness.

KDD is the oldest of three frameworks. CRISP-DM and SEMMA seem to be the practical implementation of the KDD process. CRISP-DM is more complete, as the iterative flow of the knowledge across and between phases has been clearly defined. Also, it covers all areas of building a reliable ML system from a business world perspective. In SEMMA's sample stage it's important that you have a true understanding of all aspects of the business, to ensure the sampled data retains maximum information. However, drastic innovation in the recent past has led to reduced cost for data storage and computational power, which enables us to apply ML algorithms on the entire data efficiently, almost removing the need for sampling.

We can see that generally the core phases are covered by all three frameworks and there is not a huge difference between them (Figure 2-12). Overall, these processes guide us about how data mining techniques can be applied to practical scenarios. In general, most of the researchers and data mining experts follow the KDD and CRISP-DM process model because it is more complete and accurate. I personally recommend following CRISP-DM for usage in a business environment, as it provides coverage of end to end business activity and the life cycle of building an ML system.

Summary of data mining frameworks

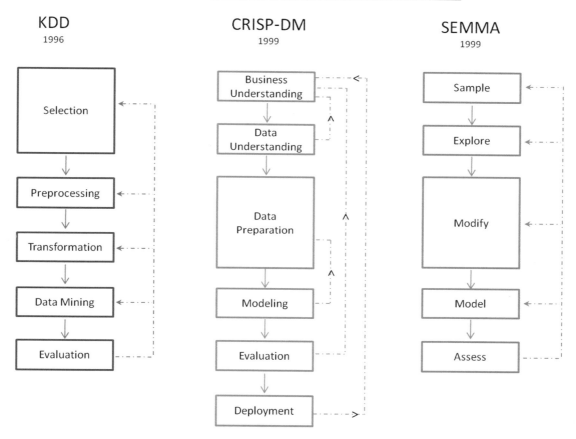

Figure 2-12. *Summary of data mining frameworks*

Machine Learning Python Packages

There is a rich number of open source libraries available to facilitate practical ML. These are mainly known as scientific Python libraries and generally put to use when performing elementary ML tasks. At a high level we can divide these libraries into data analysis and core ML libraries, based on their usage/purpose.

Data analysis: These are the set of packages that provide us the mathematics and scientific functionalities that are essential to perform the data preprocessing and transformation.

Core machine learning packages: These are the set of packages that provide us all the necessary ML algorithms and functionalities that can be applied on a given dataset to extract the patterns.

Data Analysis Packages

There are four key packages that are most widely used for data analysis:

- NumPy

- SciPy

- Matplotlib

- Pandas

Pandas, NumPy, and Matplotlib play a major role and have a scope of usage in almost all data analysis tasks (Figure 2-13). So in this chapter, we'll focus on usage or concepts relevant to these three packages as much as possible. SciPy supplements the NumPy library and has a variety of key high-level science and engineering modules; however, the usage of these functions largely depends on the use case. So we'll the touch on or highlight some of the useful functionalities in coming chapters where possible.

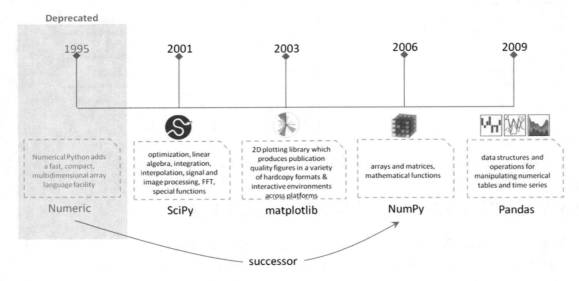

Figure 2-13. *Data analysis packages*

Note   For conciseness, we'll only be covering the key concepts within each of the libraries with a brief introduction and code implementation. You can always refer to the official user documents for these packages, which have been well designed by the developer community to cover a lot more in depth.

NumPy

NumPy is the core library for scientific computing in Python. It provides a high-performance, multidimensional array object and tools for working with these arrays. It's a successor to the Numeric package. In 2005, Travis Oliphant created NumPy by incorporating features of the competing Numarray into Numeric, with extensive modifications. I think the concepts and the code examples to a great extent have been explained in the simplest form in his book *Guide to NumPy*. Here we'll only be looking at some of the key NumPy concepts that are a must, or good to know in relevance to ML.

Array

A NumPy array is a collection of similar data type values and is indexed by a tuple of non-negative numbers. The rank of the array is the number of dimensions, and the shape of an array is a tuple of numbers giving the size of the array along each dimension.

We can initialize NumPy arrays from nested Python lists, and access elements using square brackets (Listing 2-1).

Listing 2-1. Example Code for Initializing NumPy Array

```
import numpy as np

# Create a rank 1 array
a = np.array([0, 1, 2])
print (type(a))

# this will print (the dimension of the array
print (a.shape)
print (a[0])
print (a[1])
print (a[2])

# Change an element of the array
a[0] = 5
print (a)
# ----output-----
<class 'numpy.ndarray'>
(3,)
```

```
0
1
2
[5 1 2]

# Create a rank 2 array
b = np.array([[0,1,2],[3,4,5]])
print (b.shape)
print (b)
print (b[0, 0], b[0, 1], b[1, 0])
----output-----
(2, 3)
[[0 1 2]
 [3 4 5]]
0 1 3
```

Creating NumPy Array

NumPy also provides many built-in functions to create arrays. The best way to learn this is through examples (Listing 2-2), so let's jump into the code.

Listing 2-2. Creating NumPy Array

```
# Create a 3x3 array of all zeros
a = np.zeros((3,3))
print (a)
----- output -----
[[ 0.  0.  0.]
 [ 0.  0.  0.]
 [ 0.  0.  0.]]

# Create a 2x2 array of all ones
b = np.ones((2,2))
print (b)
---- output ----
[[ 1.  1.]
 [ 1.  1.]]
```

```
# Create a 3x3 constant array
c = np.full((3,3), 7)
print (c)
---- output ----
[[7 7 7]
 [7 7 7]
 [7 7 7]]
```

```
# Create a 3x3 array filled with random values
d = np.random.random((3,3))
print (d)
---- output ----
[[0.67920283 0.54527415 0.89605908]
 [0.73966284 0.42214293 0.10170252]
 [0.26798364 0.07364324 0.260853   ]]
```

```
# Create a 3x3 identity matrix
e = np.eye(3)
print (e)
---- output ----
[[ 1.  0.  0.]
 [ 0.  1.  0.]
 [ 0.  0.  1.]]
```

```
# convert list to array
f = np.array([2, 3, 1, 0])
print (f)
---- output ----
[2 3 1 0]
```

```
# arange() will create arrays with regularly incrementing values
g = np.arange(20)
print (g)
---- output ----
[ 0  1  2  3  4  5  6  7  8  9 10 11 12 13 14 15 16 17 18 19]
```

```
# note mix of tuple and lists
h = np.array([[0, 1,2.0],[0,0,0],(1+1j,3.,2.)])
print (h)
---- output ----
[[ 0.+0.j  1.+0.j  2.+0.j]
 [ 0.+0.j  0.+0.j  0.+0.j]
 [ 1.+1.j  3.+0.j  2.+0.j]]

# create an array of range with float data type
i = np.arange(1, 8, dtype=np.float)
print (i)
---- output ----
[ 1.  2.  3.  4.  5.  6.  7.]

# linspace() will create arrays with a specified number of items which are
# spaced equally between the specified beginning and end values
j = np.linspace(2., 4., 5)
print (j)
---- output ----
[ 2.   2.5 3.   3.5 4. ]

# indices() will create a set of arrays stacked as a one-higher
# dimensioned array, one per dimension with each representing variation
# in that dimension
k = np.indices((2,2))
print (k)
---- output ----
[[[0 0]
  [1 1]]

 [[0 1]
  [0 1]]]
```

Data Types

The array is a collection of items of the same data type. NumPy supports and provided a built-in function to construct an array with an optional argument to explicitly specify the required data type (Listing 2-3).

Listing 2-3. NumPy Data Types

```
# Let numpy choose the data type
x = np.array([0, 1])
y = np.array([2.0, 3.0])

# Force a particular data type
z = np.array([5, 6], dtype=np.int64)

print (x.dtype, y.dtype, z.dtype)
---- output ----
int32 float64 int64
```

Array Indexing

NumPy offers several ways to index into arrays. Standard Python x[obj] syntax can be used to index NumPyarray, where x is the array and obj the selection.

There are three kinds of indexing available:

- Field access
- Basic slicing
- Advanced indexing

Field Access

If the ndarray object is a structured array, the fields of the array can be accessed by indexing the array with strings, dictionary-like. Indexing x['field-name'] returns a new view to the array that is of the same shape as x, except when the field is a subarray, but of data type x.dtype['field-name'] and contains only the part of the data in the specified field (Listing 2-4).

Listing 2-4. Field Access

```
x = np.zeros((3,3), dtype=[('a', np.int32), ('b', np.float64, (3,3))])
print ("x['a'].shape: ",x['a'].shape)
print ("x['a'].dtype: ", x['a'].dtype)
print ("x['b'].shape: ", x['b'].shape)
print ("x['b'].dtype: ", x['b'].dtype)
----output-----
```

```
x['a'].shape:  (3, 3)
x['a'].dtype:  int32
x['b'].shape:  (3, 3, 3, 3)
x['b'].dtype:  float64
```

Basic Slicing

NumPy arrays can be sliced, similar to lists. You must specify a slice for each dimension of the array, as the arrays may be multidimensional.

The basic slice syntax is i: j: k, where i is the starting index, j is the stopping index, and k is the step and k is not equal to 0. This selects the m elements in the corresponding dimension, with index values i, i + k, ...,i + (m - 1) k, where m = q + (r not equal to 0) and q and r are the quotient and remainder obtained by dividing j - i by k: j - i = q k + r, so that i + (m - 1) k < j. Refer to Listings 2-5 to 2-10 for example codes on basic slicing.

Listing 2-5. Basic Slicing

```
x = np.array([5, 6, 7, 8, 9])
x[1:7:2]
---- output ----
array([6, 8])
```

Negative k makes stepping go toward smaller indices. Negative i and j are interpreted as n + i and n + j, where n is the number of elements in the corresponding dimension.

Listing 2-6. Basic Slicing (continued)

```
print (x[-2:5])
print (x[-1:1:-1])
# ---- output ----
[8 9]
[9 8 7]
```

If n is the number of items in the dimension being sliced, if i is not given then it defaults to 0 for k > 0 and n - 1 for k < 0. If j is not given it defaults to n for k > 0 and -1 for k < 0. If k is not given it defaults to 1. Note that :: is the same as : and means select all indices along this axis.

Listing 2-7. Basic Slicing (continued)

```
x[4:]
# ---- output ----
array([9])
```

If the number of objects in the selection tuple is less than N, then : is assumed for any subsequent dimensions.

Listing 2-8. Basic Slicing (continued)

```
y = np.array([[[1],[2],[3]], [[4],[5],[6]]])
print ("Shape of y: ", y.shape)
y[1:3]
# ---- output ----
Shape of y:  (2, 3, 1)
```

Ellipsis expands to the number of : objects needed to make a selection tuple of the same length as x.ndim. There may only be a single ellipsis present.

Listing 2-9. Basic Slicing (continued)

```
x[...,0]
---- output ----
array(5)

# Create a rank 2 array with shape (3, 4)
a = np.array([[5,6,7,8], [1,2,3,4], [9,10,11,12]])
print ("Array a:", a)

# Use slicing to pull out the subarray consisting of the first 2 rows
# and columns 1 and 2; b is the following array of shape (2, 2):
# [[2 3]
#  [6 7]]
b = a[:2, 1:3]
print ("Array b:", b)
---- output ----
Array a:  [[ 5  6  7  8]
 [ 1  2  3  4]
```

```
[  9 10 11 12]]
Array b:  [[6 7]
 [2 3]]
```

A slice of an array is a view of the same data, so modifying it will modify the original array.

Listing 2-10. Basic Slicing (continued)

```
print (a[0, 1])
b[0, 0] = 77
print(a[0, 1])
---- output ----
6
77
```

A middle row array can be accessed in two ways: 1) slices along with integer indexing will result in an array of lower rank and 2) using only slices will result in the same rank array.

Example code:

```
# Create the following rank 2 array with shape (3, 4)
a = np.array([[1,2,3,4], [5,6,7,8], [9,10,11,12]])

row_r1 = a[1,:]# Rank 1 view of the second row of a
row_r2 = a[1:2,:]# Rank 2 view of the second row of a
print (row_r1, row_r1.shape)
print (row_r2, row_r2.shape)
---- output ----
[5 6 7 8] (4,)
[[5 6 7 8]] (1, 4)

# We can make the same distinction when accessing columns of an array:
col_r1 = a[:, 1]
col_r2 = a[:, 1:2]
print (col_r1, col_r1.shape)
print (col_r2, col_r2.shape)
---- output ----
```

```
[ 2  6 10] (3,)
[[ 2]
 [ 6]
 [10]] (3, 1)
```

Advanced Indexing

There are two kinds of advanced indexing: integer array and Boolean array.

Integer array indexing allows you to convert random arrays into another new array, as shown in Listing 2-11.

Listing 2-11. Advanced Indexing

```
a = np.array([[1,2], [3, 4]])

# An example of integer array indexing.
# The returned array will have shape (2,) and
print (a[[0, 1], [0, 1]])

# The preceding example of integer array indexing is equivalent to this:
print (np.array([a[0, 0], a[1, 1]]))
--- output ----
[1 4]
[1 4]

# When using integer array indexing, you can reuse the same
# element from the source array:
print (a[[0, 0], [1, 1]])

# Equivalent to the previous integer array indexing example
print (np.array([a[0, 1], a[0, 1]]))
---- output ----
[2 2]
[2 2]
```

Boolean array indexing is useful to pick a random element from an array, which is often used for filtering elements that satisfy a given condition (Listing 2-12).

Listing 2-12. Boolean Array Indexing

```
a = np.array([[1,2], [3, 4], [5, 6]])
# Find the elements of a that are bigger than 2
print (a > 2)

# to get the actual value
print (a[a > 2])
---- output ----
[[False False]
 [ True  True]
 [ True  True]]
[3 4 5 6]
```

Array Math

Basic mathematical functions are available as operator and also as functions in NumPy. It operates elementwise on the array (Listing 2-13).

Listing 2-13. Array Math

```
import numpy as np

x=np.array([[1,2],[3,4],[5,6]])
y=np.array([[7,8],[9,10],[11,12]])

# Elementwise sum; both produce the array
print (x+y)
print (np.add(x, y))
---- output ----
[[ 8 10]
 [12 14]
 [16 18]]
[[ 8 10]
 [12 14]
 [16 18]]

# Elementwise difference; both produce the array
print(x-y)
```

```
print (np.subtract(x, y))
---- output ----
[[-6 -6]
 [-6 -6]
 [-6 -6]]
[[-6 -6]
 [-6 -6]
 [-6 -6]]

# Elementwise product; both produce the array
print (x*y)
print (np.multiply(x, y))
---- output ----
[[ 7 16]
 [27 40]
 [55 72]]
[[ 7 16]
 [27 40]
 [55 72]]

# Elementwise division; both produce the array
print (x/y)
print (np.divide(x, y))
---- output ----
[[0.14285714 0.25      ]
 [0.33333333 0.4       ]
 [0.45454545 0.5       ]]
[[0.14285714 0.25      ]
 [0.33333333 0.4       ]
 [0.45454545 0.5       ]]

# Elementwise square root; produces the array
print(np.sqrt(x))
---- output ----
[[1.         1.41421356]
 [1.73205081 2.        ]
 [2.23606798 2.44948974]]
```

We can use the "dot" function to calculate inner products of vectors or to multiply matrices or multiply vector by matrix, as shown in the code example of Listing 2-14.

Listing 2-14. Array Math (continued)

```
x = np.array([[1,2],[3,4]])
y = np.array([[5,6],[7,8]])

a = np.array([9,10])
b = np.array([11, 12])

# Inner product of vectors; both produce 219
print (a.dot(b))
print (np.dot(a, b))
---- output ----
219
219

# Matrix / vector product; both produce the rank 1 array [29 67]
print (x.dot(a))
print (np.dot(x, a))
---- output ----
[29 67]
[29 67]

# Matrix / matrix product; both produce the rank 2 array
print (x.dot(y))
print (np.dot(x, y))
---- output ----
[[19 22]
 [43 50]]
[[19 22]
 [43 50]]
```

NumPy provides many useful functions for performing computations on arrays. One of the most useful is "sum"; example code is shown in Listing 2-15.

Listing 2-15. Sum Function

```
x = np.array([[1,2],[3,4]])

# Compute sum of all elements
print (np.sum(x))
# Compute sum of each column
print (np.sum(x, axis=0))
# Compute sum of each row
print (np.sum(x, axis=1))
---- output ----
10
[4 6]
[3 7]
```

Transpose is one of the common operations often performed on the matrix, which can be achieved using the T attribute of an array object. Refer to Listing 2-16 for code example.

Listing 2-16. Transpose Function

```
x = np.array([[1,2],[3,4]])
print (x)
print (x.T)
---- output ----
[[1 2]
 [3 4]]
[[1 3]
 [2 4]]

# Note that taking the transpose of a rank 1 array does nothing:
v = np.array([1,2,3])
print (v)
print (v.T)
---- output ----
[1 2 3]
[1 2 3]
```

Broadcasting

Broadcasting enables arithmetic operations to be performed between differently shaped arrays. Let's look at a simple example (Listing 2-17) of adding a constant vector to each row of the matrix.

Listing 2-17. Broadcasting

```
# create a matrix
a = np.array([[1,2,3], [4,5,6], [7,8,9]])
# create a vector
v = np.array([1, 0, 1])

# Create an empty matrix with the same shape as a
b = np.empty_like(a)

# Add the vector v to each row of the matrix x with an explicit loop
for i in range(3):
    b[i, :] = a[i, :] + v

print (b)
---- output ----
[[ 2  2  4]
 [ 5  5  7]
 [ 8  8 10]]
```

Performing the preceding operation on a large matrix through the loop in Python could be slow. Let's look at an alternative approach shown in Listing 2-18.

Listing 2-18. Broadcasting for Large Matrix

```
# Stack 3 copies of v on top of each other
vv = np.tile(v, (3, 1))
print (vv)
---- output ----
[[1 0 1]
 [1 0 1]
 [1 0 1]]
```

```
# Add x and vv elementwise
b = a + vv
print (b)
---- output ----
[[ 2  2  4]
 [ 5  5  7]
 [ 8  8 10]]
```

Now let's see how the preceding can be achieved using NumPy broadcasting in the example code in Listing 2-19.

Listing 2-19. Broadcasting Using NumPy

```
a = np.array([[1,2,3], [4,5,6], [7,8,9]])
v = np.array([1, 0, 1])

# Add v to each row of a using broadcasting
b = a + v
print (b)
---- output ----
[[ 2  2  4]
 [ 5  5  7]
 [ 8  8 10]]
```

Now let's look at some applications of broadcasting (Listing 2-20).

Listing 2-20. Appliclations of Broadcasting

```
# Compute outer product of vectors
# v has shape (3,)
v = np.array([1,2,3])
# w has shape (2,)
w = np.array([4,5])
# To compute an outer product, we first reshape v to be a column
# vector of shape (3, 1); we can then broadcast it against w to yield
# an output of shape (3, 2), which is the outer product of v and w:
```

```
print (np.reshape(v, (3, 1)) * w)
---- output ----
[[ 4  5]
 [ 8 10]
 [12 15]]

# Add a vector to each row of a matrix
x = np.array([[1,2,3], [4,5,6]])
# x has shape (2, 3) and v has shape (3,) so they broadcast to (2, 3)

print (x + v)

---- output ----
[[2 4 6]
 [5 7 9]]

# Add a vector to each column of a matrix
# x has shape (2, 3) and w has shape (2,).
# If we transpose x then it has shape (3, 2) and can be broadcast
# against w to yield a result of shape (3, 2); transposing this result
# yields the final result of shape (2, 3) which is the matrix x with
# the vector w added to each column

print ((x.T + w).T)
---- output ----
[[ 5  6  7]
 [ 9 10 11]]

# Another solution is to reshape w to be a row vector of shape (2, 1);
# we can then broadcast it directly against x to produce the same
# output.
print (x + np.reshape(w, (2, 1)))
---- output ----
[[ 5  6  7]
 [ 9 10 11]]

# Multiply a matrix by a constant:
# x has shape (2, 3). Numpy treats scalars as arrays of shape ();
# these can be broadcast together to shape (2, 3), producing the
```

```
# following array:
print (x * 2)
---- output ----
[[ 2  4  6]
 [ 8 10 12]]
```

Broadcasting typically makes your code more concise and faster, so you should strive to use it where possible.

Pandas

Python has always been great for data munging; however, it was not great for analysis compared with databases using SQL or Excel or R data frames. Pandas is an open source Python package providing fast, flexible, and expressive data structures designed to make working with "relational" or "labeled" data both easy and intuitive. Pandas was developed by Wes McKinney in 2008 while at AQR Capital Management, out of the need for a high performance, flexible tool to perform quantitative analysis of financial data. Before leaving AQR he was able to convince management to allow him to open source the library.

Pandas is well suited for tabular data with heterogeneously typed columns, as in an SQL table or Excel spreadsheet.

Data Structures

Pandas introduced two new data structures to Python—Series, and DataFrame—both of which are built on top of NumPy (this means they're fast).

Series

This is a one-dimensional object similar to a column in a spreadsheet or SQL table. By default, each item will be assigned an index label from 0 to N (Listing 2-21).

Listing 2-21. Creating a Pandas Series

```
import pandas as pd

# creating a series by passing a list of values, and a custom index label.
# Note that the labeled index reference for each row and it can have
duplicate values
```

```
s = pd.Series([1,2,3,np.nan,5,6], index=['A','B','C','D','E','F'])
print (s)
---- output ----
A    1.0
B    2.0
C    3.0
D    NaN
E    5.0
F    6.0
dtype: float64
```

DataFrame

It is a two-dimensional object similar to a spreadsheet or an SQL table. This is the most commonly used Pandas object (Listing 2-22).

Listing 2-22. Creating a Pandas DataFrame

```
data = {'Gender': ['F', 'M', 'M'],
        'Emp_ID': ['E01', 'E02', 'E03'],
        'Age': [25, 27, 25]}

# We want to order the columns, so lets specify in columns parameter
df = pd.DataFrame(data, columns=['Emp_ID','Gender', 'Age'])
df
---- output ----
       Emp_ID       Gender       Age
0      E01          F            25
1      E02          M            27
2      E03          M            25
```

Reading and Writing Data

We'll see three commonly used file formats: csv, text file and Excel (Listing 2-23).

Listing 2-23. Reading/Writing Data from csv, text, Excel

```
# Reading
df=pd.read_csv('Data/mtcars.csv')                    # from csv
```

```
df=pd.read_csv('Data/mtcars.txt', sep='\t')    # from text file
df=pd.read_excel('Data/mtcars.xlsx','Sheet2') # from Excel

# reading from multiple sheets of same Excel into different dataframes
xlsx = pd.ExcelFile('file_name.xls')
sheet1_df = pd.read_excel(xlsx, 'Sheet1')
sheet2_df = pd.read_excel(xlsx, 'Sheet2')

# writing
# index = False parameter will not write the index values, default is True
df.to_csv('Data/mtcars_new.csv', index=False)
df.to_csv('Data/mtcars_new.txt', sep='\t', index=False)
df.to_excel('Data/mtcars_new.xlsx',sheet_name='Sheet1', index = False)
```

Note   By default, Write will overwrite any existing file with the same name.

Basic Statistics Summary

Pandas has some built-in functions to help us to get a better understanding of data, using basic statistical summary methods (Listing 2-24).

describe() will return the quick stats such as count, mean, std (standard deviation), min, first quartile, median, third quartile, and max on each column of the DataFrame.

Listing 2-24. Basic Statistics on DataFrame

```
df = pd.read_csv('Data/iris.csv')
df.describe()
---- output ----
      Sepal.Length Sepal.Width  Petal.Length Petal.Width
count 150.000000   150.000000   150.000000   150.000000
mean  5.843333     3.057333     3.758000     1.199333
std   0.828066     0.435866     1.765298     0.762238
min   4.300000     2.000000     1.000000     0.100000
25%   5.100000     2.800000     1.600000     0.300000
50%   5.800000     3.000000     4.350000     1.300000
75%   6.400000     3.300000     5.100000     1.800000
max   7.900000     4.400000     6.900000     2.500000
```

cov() covariance indicates how two variables are related. A positive covariance means the variables are positively related, whereas a negative covariance means the variables are inversely related. The drawback of covariance is that it does not tell you the degree of a positive or negative relation (Listing 2-25).

Listing 2-25. Creating Covariance on DataFrame

```
df = pd.read_csv('Data/iris.csv')
df.cov()
---- output ----
Sepal.Length  Sepal.Width  Petal.Length  Petal.Width
Sepal.Length 0.685694     -0.042434      1.274315      0.516271
Sepal.Width -0.042434      0.189979     -0.329656     -0.121639
Petal.Length 1.274315     -0.329656      3.116278      1.295609
Petal.Width  0.516271     -0.121639      1.295609      0.581006
```

corr() correlation is another way to determine how two variables are related. In addition to telling you whether variables are positively or inversely related, the correlation also tells you the degree to which the variables tend to move together. When you say that two items correlate, you are saying that the change in one item effects a change in another item. You will always talk about correlation as a range between -1 and 1. In the following example code, petal length is 87% positively related to sepal length; that means a change in petal length results in a positive 87% change to sepal length and vice versa (Listing 2-26).

Listing 2-26. Creating Correlation Matrix on DataFrame

```
df = pd.read_csv('Data/iris.csv')
df.corr()
----output----
             Sepal.Length Sepal.Width  Petal.Length Petal.Width
Sepal.Length 1.000000     -0.117570     0.871754     0.817941
Sepal.Width  -0.117570     1.000000    -0.428440    -0.366126
Petal.Length 0.871754     -0.428440     1.000000     0.962865
Petal.Width  0.817941     -0.366126     0.962865     1.000000
```

Viewing Data

Pandas DataFrame comes with built-in functions to view the contained data (Table 2-2).

Table 2-2. *Pandas View Function*

Describe	Syntax
Looking at the top n recordsdefault n value is 5 if not specified	df.head(n=2)
Looking at the bottom n records	df.tail()
Get column names	df.columns
Get column data types	df.dtypes
Get DataFrame index	df.index
Get unique values	df[column_name].unique()
Get values	df.values
Sort DataFrame	df.sort_values(by =['Column1', 'Column2'], ascending=[True,True])
Select/view by column name	df[column_name]
Select/view by row number	df[0:3]
Selection by index	df.loc[0:3] # index 0 to 3 df.loc[0:3,['column1','column2']] # index 0 to 3 for specific columns
Selection by position	df.iloc[0:2] # using range, first 2 rows df.iloc[2,3,6] # specific position df.iloc[0:2,0:2] # first 2 rows and first 2 columns
Selection without it being in the index	print (df.iat[1,1]) # value from first row and first column print (df.iloc[:,2]) # all rows of the column at 2nd position
Faster alternative to iloc to get scalar values	print (df.iloc[1,1])
Transpose DataFrame	df.T
Filter DataFrame based on value condition for one column	df[df['column_name'] > 7.5]
Filter DataFrame based on a value condition on one column	df[df['column_name'].isin(['condition_value1', 'condition_value2'])]

(continued)

Table 2-2. (*continued*)

Describe	Syntax
Filter based on multiple conditions on multiple columns using AND operator	df[(df['column1']>7.5) & (df['column2']>3)]
Filter based on multiple conditions on multiple columns using OR operator	df[(df['column1']>7.5) | (df['column2']>3)]

Basic Operations

Pandas comes with a rich set of built-in functions for basic operations (Table 2-3).

Table 2-3. *Pandas Basic Operations*

Description	Syntax
Convert string to date series	pd.to_datetime(pd. Series(['2017-04-01','2017-04-02','2017-04-03']))
Rename a specific column name	df.rename(columns={'old_columnname':'new_ columnname'}, inplace=True)
Rename all column names of DataFrame	df.columns = ['col1_new_name','col2_new_name'....]
Flag duplicates	df.duplicated()
Drop duplicates	df = df.drop_duplicates()
Drop duplicates in specific column	df.drop_duplicates(['column_name'])
Drop duplicates in specific column, but retain the first or last observation in duplicate set	df.drop_duplicates(['column_name'], keep = 'first') # change to last for retaining last obs of duplicate
Creating new column from existing column	df['new_column_name'] = df['existing_column_name'] + 5
Creating new column from elements of two columns	df['new_column_name'] = df['existing_column1'] + '_' + df['existing_column2']
Adding a list a new column to DataFrame	df['new_column_name'] = pd.Series(mylist)
Drop missing rows and columns having missing values	df.dropna()
Replaces all missing values with 0 (or you can use any int or str)	df.fillna(value=0)

(continued)

Table 2-3. (*continued*)

Description	Syntax

Replace missing values with last valid observation (useful in time series data). For example, temperature does not change drastically compared with previous observation. So, better approach to fill NA is to forward or backward fill rather than mean. There are mainly two methods available
 	
  1) 'pad' / 'ffill' - forward fill
  2) 'bfill' / 'backfill' - backward fill
 	
Limit: If method is specified, this is the maximum number of consecutive NaN values to forward/backward fill
	df.fillna(method='ffill', inplace=True, limit = 1)

Check missing value condition and return a boolean value of True or False for each cell	pd.isnull(df)
Replace all missing values for a given column with its mean	mean=df['column_name'].mean(); df['column_name'].fillna(mean)
Return mean for each column	df.mean()
Return max for each column	df.max()
Return min for each column	df.min()
Return sum for each column	df.sum()
Return count for each column	df.count()
Return cumulative sum for each column	df.cumsum()
Applies a function along an axis of the DataFrame	df.apply(np.cumsum)
Iterate over each element of a series and perform desired action	df['column_name'].map(lambda x: 1+x) # this iterates over the column and adds value 1 to each element

Apply a function to each element of DataFrame	func = lambda x: x + 1 # function to add a constant 1 to each element of DataFrame
	df.applymap(func)

Merge/Join

Pandas provides various facilities for easily combining together Series, DataFrame, and Panel objects with various kinds of set logic for the indexes and relational algebra functionality in the case of join/merge-type operations (Listing 2-27).

Listing 2-27. Concat or Append Operation

```python
data = {
        'emp_id': ['1', '2', '3', '4', '5'],
        'first_name': ['Jason', 'Andy', 'Allen', 'Alice', 'Amy'],
        'last_name': ['Larkin', 'Jacob', 'A', 'AA', 'Jackson']}
df_1 = pd.DataFrame(data, columns = ['emp_id', 'first_name', 'last_name'])

data = {
        'emp_id': ['4', '5', '6', '7'],
        'first_name': ['Brian', 'Shize', 'Kim', 'Jose'],
        'last_name': ['Alexander', 'Suma', 'Mike', 'G']}
df_2 = pd.DataFrame(data, columns = ['emp_id', 'first_name', 'last_name'])

# Usingconcat
df = pd.concat([df_1, df_2])
print (df)

# or

# Using append
print (df_1.append(df_2))

# Join the two DataFrames along columns
pd.concat([df_1, df_2], axis=1)

---- output ----
# Table df_1
 emp_idfirst_namelast_name
0    1      Jason      Larkin
1    2      Andy       Jacob
2    3      Allen          A
3    4      Alice         AA
4    5      Amy       Jackson
```

```
# Table df_2
  emp_idfirst_namelast_name
0      4        Brian   Alexander
1      5        Shize       Suma
2      6          Kim       Mike
3      7         Jose          G

# concated table
   emp_idfirst_namelast_name
0      1        Jason      Larkin
1      2         Andy       Jacob
2      3        Allen           A
3      4        Alice          AA
4      5          Amy     Jackson
0      4        Brian   Alexander
1      5        Shize        Suma
2      6          Kim        Mike
3      7         Jose           G

# concated along columns
  emp_idfirst_namelast_nameemp_idfirst_namelast_name
0      1        Jason      Larkin      4       Brian   Alexander
1      2         Andy       Jacob      5       Shize        Suma
2      3        Allen           A      6         Kim        Mike
3      4        Alice          AA      7        Jose           G
4      5          Amy     Jackson   NaNNaNNaN
```

One of the common operations with DataFrames that we might come across is merging of two DataFrames based on a common column (Listing 2-28).

Listing 2-28. Merge Two DataFrames

```
# Merge two DataFrames based on the emp_id value
# in this case only the emp_id's present in both tables will be joined
pd.merge(df_1, df_2, on='emp_id')
```

```
---- output ----
  emp_id first_name_x last_name_x first_name_y last_name_y
0      4        Alice          AA        Brian   Alexander
1      5          Amy      Jackson        Shize        Suma
```

Join

Pandas offers SQL style merges as well. Left join produces a complete set of records from Table A, with the matching records where available in Table B. If there is no match, the right side will contain null (Listing 2-29).

Note: You can add a suffix to avoid duplicate; if not provided, it will automatically add x to Table A and y to Table B.

Listing 2-29. Left Join Two DataFrames

```python
# Left join
print(pd.merge(df_1, df_2, on='emp_id', how='left'))

# Merge while adding a suffix to duplicate column names of both table
print(pd.merge(df_1, df_2, on='emp_id', how='left', suffixes=('_left',
'_right')))
```

```
---- output ----
---- without suffix ----
  emp_id first_name_x last_name_x first_name_y last_name_y
0      1        Jason      Larkin          NaN         NaN
1      2         Andy       Jacob          NaN         NaN
2      3        Allen           A          NaN         NaN
3      4        Alice          AA        Brian   Alexander
4      5          Amy     Jackson        Shize        Suma
 ---- with suffix ----
  emp_id first_name_left last_name_left first_name_right last_name_right
0      1           Jason         Larkin              NaN             NaN
1      2            Andy          Jacob              NaN             NaN
2      3           Allen              A              NaN             NaN
3      4           Alice             AA            Brian       Alexander
4      5             Amy        Jackson            Shize            Suma
```

Right join produces a complete set of records from Table B, with the matching records where available in Table A. If there is no match, the left side will contain null (Listing 2-30).

Listing 2-30. Right Join Two DataFrames

```
# Left join
pd.merge(df_1, df_2, on='emp_id', how='right')
---- output ----
  emp_id first_name_x last_name_x first_name_y last_name_y
0      4        Alice          AA        Brian   Alexander
1      5          Amy     Jackson        Shize        Suma
2      6          NaN         NaN          Kim        Mike
3      7          NaN         NaN         Jose           G
```

Inner join is another common join operation on DataFrames. It produces only the set of records that match in both Table A and Table B (Listing 2-31).

Listing 2-31. Inner Join Two DataFrames

```
pd.merge(df_1, df_2, on='emp_id', how='inner')
 ---- output ----
  emp_id first_name_x last_name_x first_name_y last_name_y
0      4        Alice          AA        Brian   Alexander
1      5          Amy     Jackson        Shize        Suma
```

Outer join: Full outer join produces a set of all records in Table A and Table B, with matching records from both sides were available. If there is no match, the missing side will contain null (Listing 2-32).

Listing 2-32. Outer Join Two DataFrames

```
pd.merge(df_1, df_2, on='emp_id', how='outer')
---- output ----
  emp_id first_name_x last_name_x first_name_y last_name_y
0      1        Jason      Larkin          NaN         NaN
1      2         Andy       Jacob          NaN         NaN
2      3        Allen           A          NaN         NaN
```

3	4	Alice	AA	Brian	Alexander
4	5	Amy	Jackson	Shize	Suma
5	6	NaN	NaN	Kim	Mike
6	7	NaN	NaN	Jose	G

Grouping

Grouping involves one or more of the following steps (Listing 2-33):

1. Splitting the data into groups based on some criteria

2. Applying a function to each group independently

3. Combining the results into a data structure

Listing 2-33. Grouping Operation

```
df = pd.DataFrame({'Name' : ['jack', 'jane', 'jack', 'jane', 'jack', 'jane',
                             'jack', 'jane'],
                   'State' : ['SFO', 'SFO', 'NYK', 'CA', 'NYK', 'NYK', 'SFO', 'CA'],
                   'Grade':['A','A','B','A','C','B','C','A'],
                   'Age' : np.random.uniform(24, 50, size=8),
                   'Salary' : np.random.uniform(3000, 5000, size=8),})

# Note that the columns are ordered automatically in their alphabetic order
# for custom order please use below code
# df = pd.DataFrame(data, columns = ['Name', 'State', 'Age','Salary'])

# Find max age and salary by Name / State
# with the group by, we can use all aggregate functions such as min, max,
mean, count, cumsum
df.groupby(['Name','State']).max()

---- output ----

---- DataFrame ----
        Age Grade  Name      Salary State
0  45.364742     A  jack  3895.416684   SFO
1  48.457585     A  jane  4215.666887   SFO
2  47.742285     B  jack  4473.734783   NYK
```

```
3    35.181925    A    jane    4866.492808    CA
4    30.285309    C    jack    4874.123001    NYK
5    35.649467    B    jane    3689.269083    NYK
6    42.320776    C    jack    4317.227558    SFO
7    46.809112    A    jane    3327.306419    CA
```

```
  ----- find max age and salary by Name / State -----
                   Age Grade        Salary
Name State
jack NYK     47.742285     C    4874.123001
     SFO     45.364742     C    4317.227558
jane CA      46.809112     A    4866.492808
     NYK     35.649467     B    3689.269083
     SFO     48.457585     A    4215.666887
```

Pivot Tables

Pandas provides a function `pivot_table` to create an MS-Excel spreadsheet style pivot table. It can take the following arguments to perform the pivotal operations (Listing 2-34).

- data: DataFrame object

- values: column to aggregate

- index: row labels

- columns: column labels

- agg func: aggregation function to be used on values; the default is NumPy.mean

Listing 2-34. Pivot Tables

```
# by state and name find mean age for each grade
pd.pivot_table(df, values='Age', index=['State', 'Name'],
columns=['Grade'])
---- output ----
```

Grade		A	B	C
State	Name			
CA	jane	40.995519	NaN	NaN
NYK	jack	NaN	47.742285	30.285309
	jane	NaN	35.649467	NaN
SFO	jack	45.364742	NaN	42.320776
	jane	48.457585	NaN	NaN

Matplotlib

Matplotlib is a numerical mathematics extension of NumPy and a great package to view or present data in a pictorial or graphical format. It enables analysts and decision makers to see analytics presented visually, so they can grasp difficult concepts or identify new patterns. There are two broad ways of using pyplot (Matplotlib offers pyplot which is a collection of command style functions that make matplotlib work like MATLAB):

- Global functions
- Object-oriented

Using Global Functions

The most common and easy approach is by using global functions to build and display a global figure, using matplotlib as a global state machine (Listing 2-35). Let's look at some of the most commonly used charts:

- plt.bar – creates a bar chart
- plt.scatter – makes a scatter plot
- plt.boxplot – makes a box and whisker plot
- plt.hist – makes a histogram
- plt.plot – creates a line plot

Listing 2-35. Creating a Plot on Variables

```
import matplotlib.pyplot as plt
%matplotlib inline
```

```
# simple bar and scatter plot
x = np.arange(5)              # assume there are 5 students
y = (20, 35, 30, 35, 27)     # their test scores
plt.bar(x,y)                 # Barplot

# need to close the figure using show() or close(), if not closed any
follow-up plot commands will use the same figure.
plt.show()                   # Try commenting this an run
plt.scatter(x,y)             # scatter plot
plt.show()

# ---- output ----
```

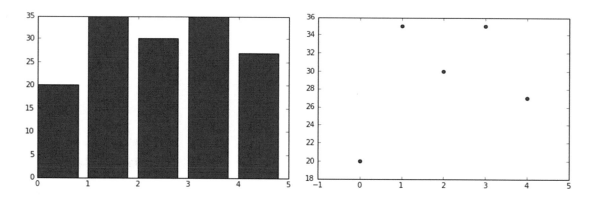

You can use the histogram, line, and boxplot directly on a DataFrame. You can see that it's very quick and does not take much coding effort (Listing 2-36).

Listing 2-36. Creating Plot on DataFrame

```
df = pd.read_csv('Data/iris.csv')    # Read sample data
df.hist()# Histogram
df.plot()                            # Line Graph
df.boxplot()                         # Box plot
#  --- histogram-------------line graph -----------box plot-------
```

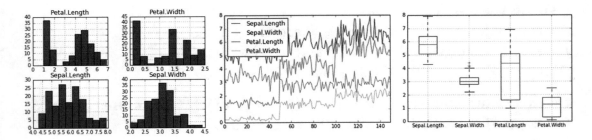

Customizing Labels

You can customize the labels to make them more meaningful, as shown in Listing 2-37.

Listing 2-37. Customize Labels

```
# generate sample data
x = np.linspace(0, 20, 1000)   #100 evenly-spaced values from 0 to 50
y = np.sin(x)

# customize axis labels
plt.plot(x, y, label = 'Sample Label')
plt.title('Sample Plot Title')                         # chart title
plt.xlabel('x axis label')                             # x axis title
plt.ylabel('y axis label')                             # y axis title
plt.grid(True)                                         # show gridlines

# add footnote
plt.figtext(0.995, 0.01, 'Footnote', ha='right', va='bottom')

# add legend, location pick the best automatically
plt.legend(loc='best', framealpha=0.5, prop={'size':'small'})

# tight_layout() can take keyword arguments of pad, w_pad and h_pad.
# these control the extra padding around the figure border and between subplots.
# The pads are specified in fraction of fontsize.
plt.tight_layout(pad=1)

# Saving chart to a file
plt.savefig('filename.png')
```

```
plt.close()  # Close the current window to allow new plot creation on
separate window / axis, alternatively we can use show()
plt.show()
```

---- output ----

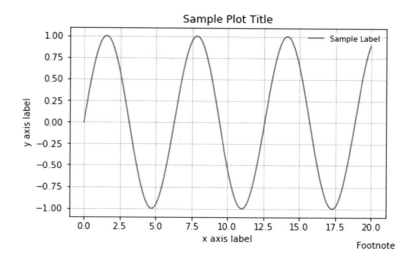

Object-Oriented

You obtain an empty Figure from a global factory and then build the plot explicitly using the methods of the Figure and the classes it contains. The Figure is the top-level container for everything on a canvas. Axes is a container class for a specific plot. A Figure may contain many axes and/or subplots. Subplots are laid out in a grid within the Figure. Axes can be placed anywhere on the Figure. We can use the subplots factory to get the Figure and all the desired axes at once. This is demonstrated in Listings 2-38 through 2-48 and Figures 2-14 through 2-16.

Listing 2-38. Object-Oriented Customization

```
fig, ax = plt.subplots()
fig,(ax1,ax2,ax3) = plt.subplots(nrows=3, ncols=1, sharex=True, figsize=(8,4))

# Iterating the Axes within a Figure
for ax in fig.get_axes():
    pass                              # do something

# ---- output ----
```

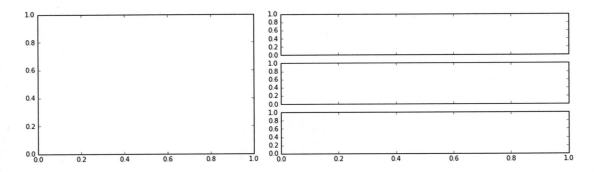

Line Plots Using ax.plot()

Here is a single plot constructed with Figure and Axes.

Listing 2-39. Single Line Plot Using ax.plot()

```
# generate sample data
x = np.linspace(0, 20, 1000)
y = np.sin(x)

fig = plt.figure(figsize=(8,4))                         # get an empty
figure and add an Axes
ax = fig.add_subplot(1,1,1)                             # row-col-num
ax.plot(x, y, 'b-', linewidth=2, label='Sample label') # line plot data on
the Axes

# add title, labels and legend, etc.
ax.set_ylabel('y axis label', fontsize=16)             # y label
ax.set_xlabel('x axis label', fontsize=16)             # x label
ax.legend(loc='best')                                  # legend
ax.grid(True)                                          # show grid
fig.suptitle('Sample Plot Title')                      # title
fig.tight_layout(pad=1)                                # tidy laytout
fig.savefig('filename.png', dpi=125)

# ---- output ----
```

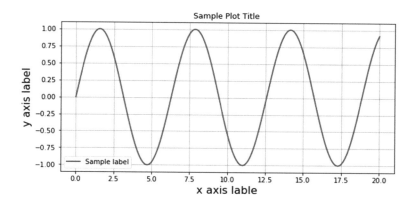

Multiple Lines on the Same Axis

You can see the code example in Listing 2-40 for plotting multiple line plots on the same axis.

Listing 2-40. Multiple Line Plot on the Same Axis

```
# get the Figure and Axes all at once
fig, ax = plt.subplots(figsize=(8,4))

x1 = np.linspace(0, 100, 20)
x2 = np.linspace(0, 100, 20)
x3 = np.linspace(0, 100, 20)
y1 = np.sin(x1)
y2 = np.cos(x2)
y3 = np.tan(x3)

ax.plot(x1, y1, label='sin')
ax.plot(x2, y2, label='cos')
ax.plot(x3, y3, label='tan')

# add grid, legend, title and save
ax.grid(True)

ax.legend(loc='best', prop={'size':'large'})

fig.suptitle('A Simple Multi Axis Line Plot')
fig.savefig('filename.png', dpi=125)

# ---- output ----
```

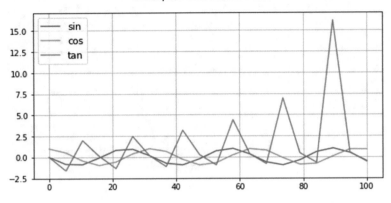

Multiple Lines on Different Axis

Refer to Listing 2-41 for the code example to draw multiple lines on different axes.

Listing 2-41. Multiple Lines on Different Axes

```
# Changing sharex to True will use the same x axis
fig, (ax1,ax2,ax3) = plt.subplots(nrows=3, ncols=1, sharex=False, sharey =
False, figsize=(8,4))

# plot some lines
x1 = np.linspace(0, 100, 20)
x2 = np.linspace(0, 100, 20)
x3 = np.linspace(0, 100, 20)
y1 = np.sin(x1)
y2 = np.cos(x2)
y3 = np.tan(x3)

ax1.plot(x1, y1, label='sin')
ax2.plot(x2, y2, label='cos')
ax3.plot(x3, y3, label='tan')

# add grid, legend, title and save
ax1.grid(True)
ax2.grid(True)
ax3.grid(True)
```

```
ax1.legend(loc='best', prop={'size':'large'})
ax2.legend(loc='best', prop={'size':'large'})
ax3.legend(loc='best', prop={'size':'large'})

fig.suptitle('A Simple Multi Axis Line Plot')
fig.savefig('filename.png', dpi=125)
# ---- output ----
```

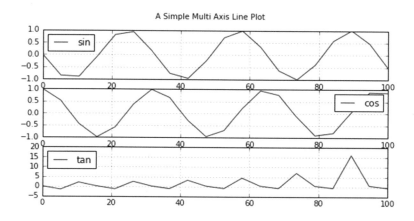

Control the Line Style and Marker Style

Refer to Listing 2-42 for the code example to understand how the line styles and marker styles can be controlled for the charts.

Listing 2-42. Line Style and Marker Style Controls

```
# get the Figure and Axes all at once
fig, ax = plt.subplots(figsize=(8,4))

# plot some lines
N = 3 # the number of lines we will plot
styles = ['-', '--', '-.', ':']
markers = list('+ox')
x = np.linspace(0, 100, 20)

for i in range(N): # add line-by-line
    y = x + x/5*i + i
    s = styles[i % len(styles)]
```

```
    m = markers[i % len(markers)]
    ax.plot(x, y, alpha = 1, label='Line '+str(i+1)+' '+s+m,
                    marker=m, linewidth=2, linestyle=s)

# add grid, legend, title and save
ax.grid(True)
ax.legend(loc='best', prop={'size':'large'})
fig.suptitle('A Simple Line Plot')
fig.savefig('filename.png', dpi=125)

# ---- output ----
```

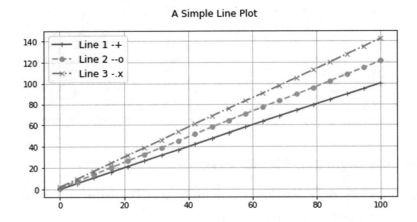

Line Style Reference

Figure 2-14 provides the summary of available matplotlib line styles.

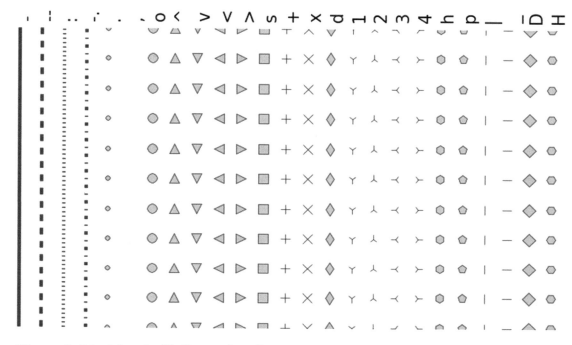

Figure 2-14. *Matplotlib line style reference*

Marker Reference

Figure 2-15 provides the summary of available matplotlib marker styles.

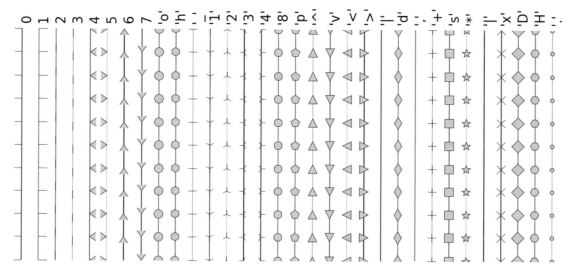

Figure 2-15. *Matplotlib marker reference*

Colormaps Reference

All color maps shown in Figure 2-16 can be reversed by appending _r. For instance, gray_r is the reverse of gray.

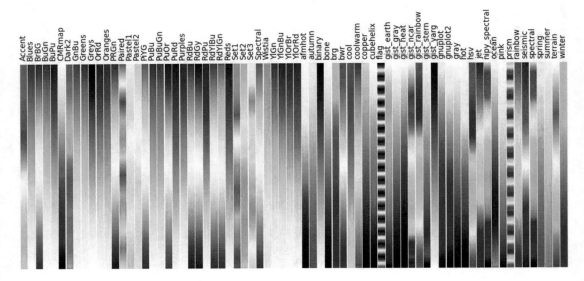

Figure 2-16. *Matplotlib colormaps reference*

Bar Plots Using ax.bar()

Refer to Listing 2-43 for the code example of bar plots using ax.bar()

Listing 2-43. Bar Plots Using ax.bar() and ax.barh()

```
# get the data
N = 4
labels = list('ABCD')
data = np.array(range(N)) + np.random.rand(N)

#plot the data
fig, ax = plt.subplots(figsize=(8, 3.5))
width = 0.5;
tickLocations = np.arange(N)
rectLocations = tickLocations-(width/2.0)
```

```
# for color either HEX value of the name of the color can be used
ax.bar(rectLocations, data, width,
       color='lightblue',
       edgecolor='#1f10ed', linewidth=4.0)

# tidy-up the plot
ax.set_xticks(ticks= tickLocations)
ax.set_xticklabels(labels)
ax.set_xlim(min(tickLocations)-0.6, max(tickLocations)+0.6)
ax.set_yticks(range(N)[1:])
ax.set_ylim((0,N))
ax.yaxis.grid(True)
ax.set_ylabel('y axis label', fontsize=8)           # y label
ax.set_xlabel('x axis label', fontsize=8)           # x label

# title and save
fig.suptitle("Bar Plot")
fig.tight_layout(pad=2)
fig.savefig('filename.png', dpi=125)
# ---- output ----
```

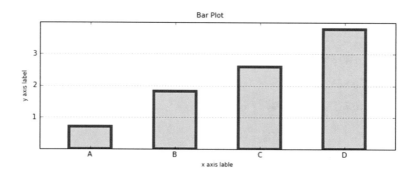

Horizontal Bar Charts Using ax.barh()

Just as tick placement needs to be managed with vertical bars, so with horizontal bars that are above the y-tick mark (Listing 2-44).

135

Listing 2-44. Horizontal Bar Charts

```
# get the data
N = 4
labels = list('ABCD')
data = np.array(range(N)) + np.random.rand(N)

#plot the data
fig, ax = plt.subplots(figsize=(8, 3.5))
width = 0.5;
tickLocations = np.arange(N)
rectLocations = tickLocations-(width/2.0)

# for color either HEX value of the name of the color can be used
ax.barh(rectLocations, data, width, color='lightblue')

# tidy-up the plot
ax.set_yticks(ticks= tickLocations)
ax.set_yticklabels(labels)
ax.set_ylim(min(tickLocations)-0.6, max(tickLocations)+0.6)
ax.xaxis.grid(True)
ax.set_ylabel('y axis label', fontsize=8)       # y label
ax.set_xlabel('x axis label', fontsize=8)       # x label

# title and save
fig.suptitle("Bar Plot")
fig.tight_layout(pad=2)
fig.savefig('filename.png', dpi=125)
# ---- output ----
```

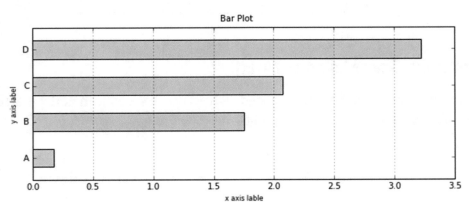

Side by Side Bar Chart

Refer to Listing 2-45 for the code example of plotting side by side bar charts.

Listing 2-45. Side by Side Bar Chart

```
# generate sample data
pre = np.array([19, 6, 11, 9])
post = np.array([15, 11, 9, 8])
labels=['Survey '+x for x in list('ABCD')]

# the plot - left then right
fig, ax = plt.subplots(figsize=(8, 3.5))
width = 0.4 # bar width
xlocs = np.arange(len(pre))
ax.bar(xlocs-width, pre, width,
       color='green', label='True')
ax.bar(xlocs, post, width,
       color='#1f10ed', label='False')

# labels, grids and title, then save
ax.set_xticks(ticks=range(len(pre)))
ax.set_xticklabels(labels)
ax.yaxis.grid(True)
ax.legend(loc='best')
ax.set_ylabel('Count')
fig.suptitle('Sample Chart')
fig.tight_layout(pad=1)
fig.savefig('filename.png', dpi=125)
# ---- output ----
```

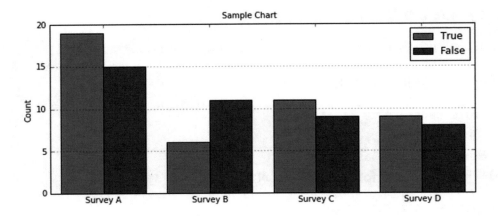

Stacked Bar Example Code

Refer to Listing 2-46 for the code example of creating stacked bar charts.

Listing 2-46. Stacked Bar Charts

```
# generate sample data
pre = np.array([19, 6, 11, 9])
post = np.array([15, 11, 9, 8])
labels=['Survey '+x for x in list('ABCD')]

# the plot - left then right
fig, ax = plt.subplots(figsize=(8, 3.5))
width = 0.4 # bar width
xlocs = np.arange(len(pre)+2)
adjlocs = xlocs[1:-1] - width/2.0
ax.bar(adjlocs, pre, width,
       color='grey', label='True')
ax.bar(adjlocs, post, width,
       color='cyan', label='False',
       bottom=pre)

# labels, grids and title, then save
ax.set_xticks(ticks=xlocs[1:-1])
ax.set_xticklabels(labels)
ax.yaxis.grid(True)
ax.legend(loc='best')
```

```
ax.set_ylabel('Count')
fig.suptitle('Sample Chart')
fig.tight_layout(pad=1)
fig.savefig('filename.png', dpi=125)
# ---- output ----
```

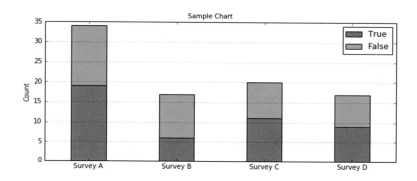

Pie Chart Using ax.pie()

Refer to Listing 2-47 for the code example to create a pie chart.

Listing 2-47. Pie Chart

```
# generate sample data
data = np.array([15,8,4])
labels = ['Feature Engineering', 'Model Tuning', 'Model Building']
explode = (0, 0.1, 0) # explode feature engineering
colrs=['cyan', 'tan', 'wheat']

# plot
fig, ax = plt.subplots(figsize=(8, 3.5))
ax.pie(data, explode=explode,
       labels=labels, autopct='%1.1f%%',
       startangle=270, colors=colrs)
ax.axis('equal') # keep it a circle

# tidy-up and save
fig.suptitle("ML Pie")
fig.savefig('filename.png', dpi=125)
# ---- output ----
```

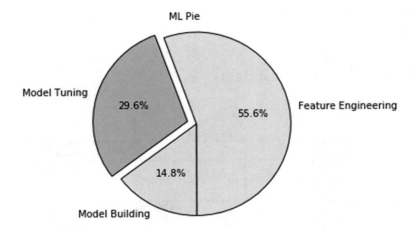

Example Code for Grid Creation

Refer to Listing 2-48 for the code example of grid creation.

Listing 2-48. Grid Creation

```
# Simple subplot grid layouts
fig = plt.figure(figsize=(8,4))
fig.text(x=0.01, y=0.01, s='Figure',color='#888888', ha='left',
va='bottom', fontsize=20)

for i in range(4):
    # fig.add_subplot(nrows, ncols, num)
    ax = fig.add_subplot(2, 2, i+1)
    ax.text(x=0.01, y=0.01, s='Subplot 2 2 '+str(i+1),  color='red',
    ha='left', va='bottom', fontsize=20)
    ax.set_xticks([]); ax.set_yticks([])
ax.set_xticks([]); ax.set_yticks([])
fig.suptitle('Subplots')
fig.savefig('filename.png', dpi=125)
# ---- output ----
```

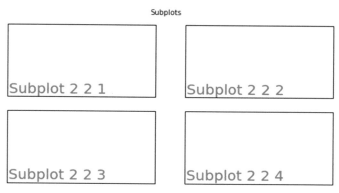

Plotting Defaults

Matplotlib uses matplotlibrc configuration files to customize all kinds of properties, which we call rc settings or rc parameters. You can control the defaults of almost every property in matplotlib, such as figure size and dpi, line width, color and style, axes, axis and grid properties, text and font properties, and so on (Listing 2-49). The location of the configuration file can be found using the following code so that you can edit it if required.

Listing 2-49. Plotting Defaults

```
# get the configuration file location
print (matplotlib.matplotlib_fname())

# get configuration current settings
print (matplotlib.rcParams)

# Change the default settings
plt.rc('figure', figsize=(8,4), dpi=125,facecolor='white',
edgecolor='white')
plt.rc('axes', facecolor='#e5e5e5',  grid=True, linewidth=1.0,
axisbelow=True)
plt.rc('grid', color='white', linestyle='-',    linewidth=2.0, alpha=1.0)
plt.rc('xtick', direction='out')
plt.rc('ytick', direction='out')
plt.rc('legend', loc='best')
```

Machine Learning Core Libraries

Python has a plethora of open source ML libraries. Table 2-4 gives a quick summary of the top 10 Python ML libraries ranked based on their number of contributors. It also shows the change in the percentage of growth in their contributors count between 2016 and 2018.

Table 2-4. *Python ML Libraries*

	Contributors				
Project Name	**2016**	**2018**	**Change %**	**License**	**Source**
Scikit-learn	732	1237	69%	BSD 3	www.github.com/scikit-learn/scikit-learn
Keras	N/A	770	N/A	MIT	https://github.com/keras-team/keras
Xgboost	N/A	338	N/A	Apache 2.0	https://github.com/dmlc/xgboost
StatsModels	N/A	167	N/A	BSD 3	https://github.com/statsmodels/statsmodels
Pylearn2	115	116	1%	BSD 3	www.github.com/lisa-lab/pylearn2
NuPIC	75	86	15%	AGPL 3	www.github.com/numenta/nupic
Nilearn	46	81	76%	BSD	www.github.com/nilearn/nilearn
PyBrain	31	32	3%	BSD 3	www.github.com/idiap/bob www.github.com/pybrain/pybrain
Pattern	20	19	-5%	BSD 3	www.github.com/clips/pattern
Fuel	29	32	10%	MIT	www.github.com/luispedro/milk www.github.com/mila-udem/fuel

Note: 2016 numbers are based on KDNuggets news

Scikit-learn is the most popular and widely used ML library. It is built on top of SciPy and features a rich number of supervised and unsupervised learning algorithms.

We'll learn more about different algorithms of Scikit-learn in detail in the next chapter.

Summary

With this, we have reached the end of this chapter. We have learned what machine learning is and where it fits in the wider AI family. We have also learned about the different related forms/terms (such as statistics, data or business analytics, data science) that exist parallel to ML and why they exist. We have briefly understood the high-level categories of ML, and the most commonly used frameworks to build efficient ML systems. Toward the end, we learned that the ML libraries can be categorized into data analysis and core ML packages. We also looked at the key concepts and example implementation code for three important data analysis packages: NumPy, Pandas, and Matplotlib. I would like to leave you with some useful resources (Table 2-5) for your future reference, to deepen your knowledge of the data analysis packages.

Table 2-5. *Additional Resources*

Resource	Description	Mode
https://docs.scipy.org/doc/numpy/reference/	This is a quick start tutorial for NumPy and covers all the concepts in detail.	Online
http://pandas.pydata.org/pandas-docs/stable/tutorials.html	This is a guide to many Pandas tutorials, geared mainly for new users.	Online
http://matplotlib.org/users/beginner.html	Beginners guide, Pyplot tutorial	Online
Python for Data Analysis	This book is concerned with the nuts and bolts of manipulating, processing, cleaning, and crunching data in Python.	Book

## CHAPTER 3

Step 3: Fundamentals of Machine Learning

This chapter focuses on different algorithms of supervised and unsupervised machine learning (ML) using two key Python packages.

Scikit-learn: In 2007, David Cournapeau developed Scikit-learn as part of the Google summer of code project. INRIA got involved in 2010 and beta v0.1 was released to the public. Currently, there are more than 700 active contributors, and paid sponsorship from INRIA, Python Software Foundation, Google, and Tinyclues. Many of the functions of Scikit-learn are built upon the SciPy (Scientific Python) library, and it provides a great breadth of efficiently implemented, essential, supervised and unsupervised learning algorithms.

Note Scikit-learn is also known as sklearn, so these two terms are used interchangeably throughout this book.

Statsmodels: This complements the SciPy package and is one of the best packages to run regression models, as it provides an extensive list of statistics results for each estimator of the model.

Machine Learning Perspective of Data

Data is the facts and figures (can also be referred to as raw data) that we have available with respect to the business context. Data is made up of two aspects:

1. *Objects* such as people, tree, animals, etc.

2. *Attributes* that were recorded for objects such as age, size, weight, cost, etc.

© Manohar Swamynathan 2019
M. Swamynathan, *Mastering Machine Learning with Python in Six Steps*,
https://doi.org/10.1007/978-1-4842-4947-5_3

When we measure the attributes of an object, we obtain a value that varies between objects. For example, if we consider individual plants in a garden as objects, the attribute "height" will vary between them. Correspondingly different attributes vary between objects, so attributes are more collectively known as variables.

The things we measure, control, or manipulate for objects are the variables. It differs as to how well they can be measured, that is, how much measurable information their measurement scale can provide. The amount of information that can be provided by a variable is determined by its type of measurement scale.

At a high level there are two types of variables based on the type of values they can take:

1. *Continuous quantitative: Variables* can take any positive or negative numerical value within a large range. Retail sales amount and insurance claim amount are examples for a continuous variable that can take any number within a large range. These types of variables are also generally called numerical variables.

2. *Discrete or qualitative:* Variables can take only particular values. Retail store location area, state, and city are examples for the discrete variable, as it can take only one particular value for a store (here "store" is our object). These types of variables are also known as categorical variables.

Scales of Measurement

In general, variables can be measured on four different scales (nominal, ordinal, interval, and ratio). Mean, median, and mode are the way to understand the central tendency—the middle point—of data distribution. Standard deviation, variance, and range are the most commonly used dispersion measures used to understand the spread of the data.

Nominal Scale of Measurement

Data are measured at the nominal level when each case is classified into one of a number of discrete categories. This is also called categorical, that is, used only for classification. As mean is not meaningful, all that we can do is to count the number of occurrences of each type and compute proportion (number of occurrence of each type/total occurrences). Refer to Table 3-1 for nominal scale examples.

Table 3-1. *Nominal Scale Examples*

Variable Name	Example Measurement Values
Color	Red, Green, Yellow, etc.
Gender	Female, Male
Football Player's Jersey Number	1, 2, 3, 4, 5, etc.

Ordinal Scale of Measurement

Data are measured on an ordinal scale if the categories imply order. The difference between ranks is consistent in direction and authority but not magnitude. Refer to Table 3-2 for ordinal scale examples.

Table 3-2. *Ordinal Scale Examples*

Variable Name	Example Measurement Values
Military rank	Second Lieutenant, First Lieutenant, Captain, Major, Lieutenant Colonel, Colonel, etc.
Clothing size	Small, Medium, Large, Extra Large, etc.
Class rank in an exam	1, 2, 3, 4, 5, etc.

Interval Scale of Measurement

If the differences between values have meanings, the data are measured at the interval scale. Refer to Table 3-3 for interval scale examples.

Table 3-3. *Interval Scale Examples*

Variable Name	Example Measurement Values
Temperature	10, 20, 30, 40, etc.
IQ rating	85–114, 115–129, 130–144, 145–159, etc.

Ratio Scale of Measurement

Data measured on a ratio scale have differences that are meaningful and relate to some true zero points. This is the most common scale of measurement. Refer to Table 3-4 for ratio scale examples.

Table 3-4. *Ratio Scale Examples*

Variable Name	Example Measurement Values
Weight	10, 20, 30, 40, 50, 60, etc.
Height	5, 6, 7, 8, 9, etc.
Age	1, 2, 3, 4, 5, 6, 7, etc.

Table 3-5 provides a quick summary of the different key scales of measurement.

Table 3-5. *Comparison of the Different Scales of Measurement*

	Scales of Measurement			
	Nominal	**Ordinal**	**Interval**	**Ratio**
Properties	Identity	Identity Magnitude	Identity Magnitude Equal intervals	Identity Magnitude Equal intervals True zero
Mathematical Operations	Count	Rank order	Addition Subtraction	Addition Subtraction Multiplication Division
Descriptive Statistics	Mode Proportion	Mode Median Range statistics	Mode Median Range statistics Variance Standard deviation	Mode Median Range statistics Variance Standard deviation

Feature Engineering

The output or the prediction quality of any ML algorithm predominantly depends on the quality of input being passed. The process of creating appropriate data features by applying the business context is called feature engineering, and it is one of the most important aspects of building an efficient ML system. The business context here means the expression of the business problem that we are trying to address, why we are trying to solve it, and what is the expected outcome. So let's understand the fundamentals of feature engineering before proceeding to different types of ML algorithms. The logical flow of raw data to the ML algorithm is represented in Figure 3-1.

Figure 3-1. *Logical flow of data in ML model building*

Data from different sources "as-is" is the raw data and when we apply business logic to process the raw data, the outcome is information (processed data). Further insight is derived from information. The process of converting raw data into information into insight with a business context to address a particular business problem is an important aspect of feature engineering. The output of feature engineering is a clean and meaningful set of features that can be consumed by algorithms to identify patterns and build an ML model, which can further be applied on unseen data to predict the possible outcome. In order to have an efficient ML system, often feature optimization is carried out to reduce the feature dimension and retain only the important/meaningful features, which will reduce the computation time and improve prediction performance. Note that ML model building is an iterative process. Let's look at some of the common practices that are part of feature engineering.

Dealing with Missing Data

Missing data can mislead or create problems for analyzing the data. In order to avoid any such issues, you need to impute missing data. There are four most commonly used techniques for data imputation:

- *Delete*: You could simply delete the rows containing missing values. This technique is more suitable and effective when a number of missing values row count is insignificant (say <5%) compare with the overall record count. You can achieve this using Panda's dropna() function.

- *Replace with the summary*: This is probably the most commonly used imputation technique. Summarization here is the mean, mode, or median for a respective column. For continuous or quantitative variables, either mean/average or mode or median value of the respective column can be used to replace the missing values. Whereas for categorical or qualitative variables, the mode (most frequent) summation technique works better. You can achieve this using Panda's fillna() function (please refer to Chapter 2 "Pandas" section).

- *Random replace*: You can also replace the missing values with a randomly picked value from the respective column. This technique would be appropriate where missing values row count is insignificant.

- *Use a predictive model*: This is an advanced technique. Here you can train a regression model for continuous variables and a classification model for categorical variables with the available data and use the model to predict the missing values.

Handling Categorical Data

Most of the ML libraries are designed to work well with numerical variables. So categorical variables in their original form of text description can't be directly used for model building. Let's learn some of the common methods of handling categorical data, based on their number of levels.

Create a dummy variable: This is a Boolean variable that indicates the presence of a category with the value 1 and 0 for absence. You should create k-1 dummy variables, where k is the number of levels. Scikit-learn provides a useful function, One Hot Encoder, to create a dummy variable for a given categorical variable (Listing 3-1).

Listing 3-1. Creating Dummy Variables

```
import pandas as pd
from patsy import dmatrices

df = pd.DataFrame({'A': ['high', 'medium', 'low'],
                   'B': [10,20,30]},
                  index=[0, 1, 2])

print df
#----output----
         A   B
0     high  10
1   medium  20
2      low  30

# using get_dummies function of pandas package
df_with_dummies= pd.get_dummies(df, prefix='A', columns=['A'])
print (df_with_dummies)
#----output----
    B  A_high  A_low  A_medium
0  10     1.0    0.0       0.0
1  20     0.0    0.0       1.0
2  30     0.0    1.0       0.0
```

Convert to number: Another simple method is to represent the text description of each level with a number by using the Label Encoder function of Scikit-learn (Listing 3-2). If the number of levels is high (example zip code, state, etc.), then you apply the business logic to combine levels to groups. For example, zip code or state can be combined with regions; however, in this method there is a risk of losing critical information. Another method is to combine categories based on similar frequency (a new category can be high, medium, low).

Listing 3-2. Converting the Categorical Variable to Numerics

```
import pandas as pd

# using pandas package's factorize function
df['A_pd_factorized'] = pd.factorize(df['A'])[0]

# Alternatively you can use sklearn package's LabelEncoder function
from sklearn.preprocessing import LabelEncoder
le = LabelEncoder()

df['A_LabelEncoded'] = le.fit_transform(df.A)
print (df)
#----output----
        A    B  A_pd_factorized  A_LabelEncoded
0    high   10                0               0
1  medium   20                1               2
2     low   30                2               1
```

Normalizing Data

The unit or scale of measurement varies for different variables, so an analysis with the raw measurement could be artificially skewed toward the variables with higher absolute values. Bringing all the different types of variable units in the same order of magnitude thus eliminates the potential outlier measurements that would misrepresent the finding and negatively affect the accuracy of the conclusion. Two broadly used methods for rescaling data are normalization and standardization.

Normalizing data can be achieved by min-max scaling. The formula is given below, which will scale all numeric values in the range 0 to 1.

$$X_{normalized} = \frac{\left(X - X_{min}\right)}{\left(X_{max} - X_{min}\right)}$$

Note   Be sure to remove extreme outliers before applying the preceding technique, as it can skew the normal values in your data to a small interval.

The standardization technique will transform the variables to have zero mean and standard deviation of one. The formula for standardization is given as follows, and the outcome is commonly known as z-scores.

$$Z = \frac{(X - \mu)}{\sigma}$$

where μ is the mean and σ is the standard deviation.

Standardization often has been the preferred method for various analysis, as it tells us where each data point lies within its distribution and gives a rough indication of outliers. Refer to Listing 3-3 for the example code of normalization and scaling.

Listing 3-3. Normalization and Scaling

```
from sklearn import datasets
import numpy as np
from sklearn import preprocessing

iris = datasets.load_iris()
X = iris.data[:, [2, 3]]
y = iris.target

std_scale = preprocessing.StandardScaler().fit(X)
X_std = std_scale.transform(X)

minmax_scale = preprocessing.MinMaxScaler().fit(X)
X_minmax = minmax_scale.transform(X)

print('Mean before standardization: petal length={:.1f}, petal width={:.1f}'
      .format(X[:,0].mean(), X[:,1].mean()))
print('SD before standardization: petal length={:.1f}, petal width={:.1f}'
      .format(X[:,0].std(), X[:,1].std()))

print('Mean after standardization: petal length={:.1f}, petal width={:.1f}'
      .format(X_std[:,0].mean(), X_std[:,1].mean()))
print('SD after standardization: petal length={:.1f}, petal width={:.1f}'
      .format(X_std[:,0].std(), X_std[:,1].std()))

print('\nMin value before min-max scaling: patel length={:.1f}, patel
width={:.1f}'
      .format(X[:,0].min(), X[:,1].min()))
```

```
print('Max value before min-max scaling: petal length={:.1f}, petal
width={:.1f}'
      .format(X[:,0].max(), X[:,1].max()))

print('Min value after min-max scaling: patel length={:.1f}, patel width={:.1f}'
      .format(X_minmax[:,0].min(), X_minmax[:,1].min()))
print('Max value after min-max scaling: petal length={:.1f}, petal width={:.1f}'
      .format(X_minmax[:,0].max(), X_minmax[:,1].max()))
#----output----
Mean before standardization: petal length=3.8, petal width=1.2
SD before standardization: petal length=1.8, petal width=0.8
Mean after standardization: petal length=-0.0, petal width=-0.0
SD after standardization: petal length=1.0, petal width=1.0

Min value before min-max scaling: patel length=1.0, patel width=0.1
Max value before min-max scaling: petal length=6.9, petal width=2.5
Min value after min-max scaling: patel length=0.0, patel width=0.0
Max value after min-max scaling: petal length=1.0, petal width=1.0
```

Feature Construction or Generation

Machine learning algorithms give the best results only when we provide them with the best possible features for the problem that you are trying to address. Often these features have to be manually created by spending a lot of time with actual raw data and trying to understand its relationship with all other data that you have collected to address a business problem.

It means thinking about aggregating, splitting, or combining features to create new features or decomposing features. Often this part is talked about as an art form, and is the key differentiator in competitive ML.

Feature construction is manual, slow, and requires subject matter expert intervention heavily to create rich features that can be exposed to predictive modeling algorithms to produce the best results.

Summarizing the data is a fundamental technique to help us understand the data quality and issues/gaps. Figure 3-2 maps the tabular and graphical data summarization methods for different data types. Note that this mapping shows the obvious or commonly used methods and not an exhaustive list.

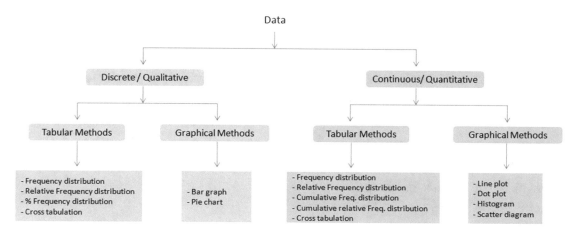

Figure 3-2. *Commonly used data summarization methods*

Exploratory Data Analysis

EDA is all about understanding your data by employing summarizing and visualizing techniques. At a high level, EDA can be performed in two ways: univariate analysis and multivariate analysis.

Let's learn to consider an example data set to learn practically. The Iris dataset is a well-known dataset used extensively in pattern recognition literature. It is hosted at the UC Irvine Machine Learning Repository. The data set contains petal length, petal width, sepal length, and sepal width measurement for three types of Iris flowers: Setosa, Versicolor, and Virginica (Figure 3-3).

Figure 3-3. *Iris Versicolor*

Univariate Analysis

Individual variables are analyzed in isolation to get a better understanding of them. Pandas provides a describe function to create summary statistics in tabular format for all variables (Listing 3-4). These statistics are very useful for the numerical type of variables, to understand any quality issues such as missing value and presence of outliers.

Listing 3-4. Univariate Analysis

```
from sklearn import datasets
import numpy as np
import pandas as pd
import matplotlib.pyplot as plt

iris = datasets.load_iris()

# Let's convert to dataframe
iris = pd.DataFrame(data= np.c_[iris['data'], iris['target']],
                    columns= iris['feature_names'] + ['species'])

# replace the values with class labels
iris.species = np.where(iris.species == 0.0, 'setosa', np.where(iris.
species==1.0,'versicolor', 'virginica'))

# let's remove spaces from column name
iris.columns = iris.columns.str.replace(' ',")
iris.describe()
#----output----
```

	sepallength(cm)	sepalwidth(cm)	petallength(cm)	petalwidth(cm)
Count	150.00	150.00	150.00	150.00
Mean	5.84	3.05	3.75	1.19
std	0.82	0.43	1.76	0.76
min	4.30	2.00	1.00	0.10
25%	5.10	2.80	1.60	0.30
50%	5.80	3.00	4.35	1.30
75%	6.40	3.30	5.10	1.80
max	7.90	4.40	6.90	2.50

The columns 'species' is categorical, so let's check the frequency distribution for each category.

```
print (iris['species'].value_counts())
#----output----
Setosa        50
versicolor    50
virginica     50
```

Pandas supports plotting functions for quick visualization of attributes. We can see from the plot that 'species' has three categories with 50 records each (Listing 3-5).

Listing 3-5. Pandas DataFrame Visualization

```
# Set the size of the plot
plt.figure(figsize=(15,8))

iris.hist()          # plot histogram
plt.suptitle("Histogram", fontsize=12) # use suptitle to add title to all sublots
plt.tight_layout(pad=1)
plt.show()

iris.boxplot()       # plot boxplot
plt.title("Bar Plot", fontsize=16)
plt.tight_layout(pad=1)
plt.show()
#----output----
```

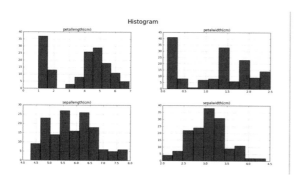

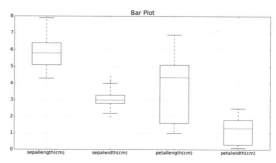

Multivariate Analysis

In multivariate analysis, you try to establish a sense of relationship of all variables with one other. Let's determine the mean of each feature by species type (Listing 3-6).

Listing 3-6. A Multivariate Analysis

```
# print the mean for each column by species
iris.groupby(by = "species").mean()

# plot for mean of each feature for each label class
iris.groupby(by = "species").mean().plot(kind="bar")

plt.title('Class vs Measurements')
plt.ylabel('mean measurement(cm)')
plt.xticks(rotation=0)  # manage the xticks rotation
plt.grid(True)

# Use bbox_to_anchor option to place the legend outside plot area to be tidy
plt.legend(loc="upper left", bbox_to_anchor=(1,1))
#----output----
```

	sepallength(cm)	sepalwidth(cm)	petallength(cm)	petalwidth(cm)
setosa	5.006	3.418	1.464	0.244
versicolor	5.936	2.770	4.260	1.326
virginica	6.588	2.974	5.552	2.026

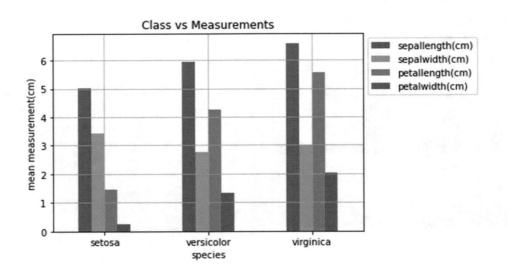

Correlation Matrix

The correlation function uses a Pearson correlation coefficient, which results in a number between -1 to 1. A strong negative relationship is indicated by a coefficient closer to -1 and a strong positive correlation is indicated by a coefficient toward 1 (Listing 3-7).

Listing 3-7. Correlation Matrix

```
# create correlation matrix
corr = iris.corr()
print(corr)

import statsmodels.api as sm
sm.graphics.plot_corr(corr, xnames=list(corr.columns))
plt.show()
#----output----
```

	sepallength(cm)	sepalwidth(cm)	petallength(cm)
sepallength(cm)	1.000000	-0.109369	0.871754
sepalwidth(cm)	-0.109369	1.000000	-0.420516
petallength(cm)	0.871754	-0.420516	1.000000
petalwidth(cm)	0.817954	-0.356544	0.962757

	petalwidth(cm)
sepallength(cm)	0.817954
sepalwidth(cm)	-0.356544
petallength(cm)	0.962757
petalwidth(cm)	1.000000

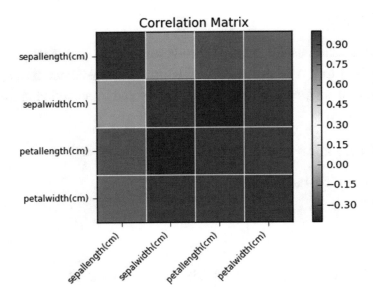

Pair Plot

You can understand the relationship attributes by looking at the distribution of the interactions of each pair of attributes. This uses a built-in function to create a matrix of scatter plots of all attributes against all attributes (Listing 3-8).

Listing 3-8. Pair Plot

```
from pandas.plotting import scatter_matrix
scatter_matrix(iris, figsize=(10, 10))

# use suptitle to add title to all sublots
plt.suptitle("Pair Plot", fontsize=20)
#----output----
```

Pair Plot

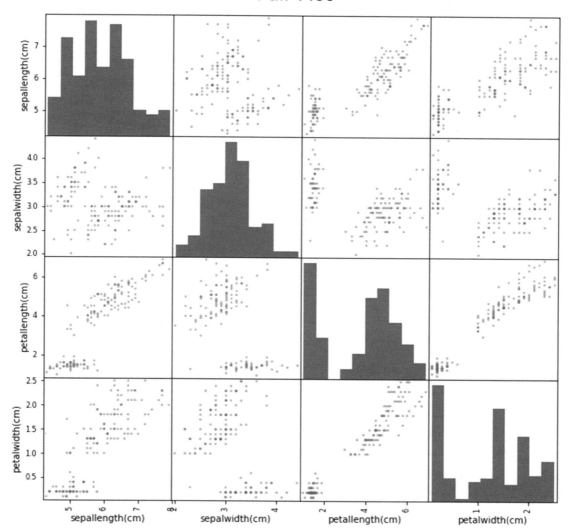

Findings from EDA

- There are no missing values.

- Sepal is longer than a petal. Sepal length ranges between 4.3 and 7.9 with an average length of 5.8, whereas petal length ranges between 1 and 6.9 with an average length of 3.7.

- Sepal is also wider than a petal. Sepal width ranges between 2 to 4.4 with an average width of 3.05, whereas petal width ranges between 0.1 to 2.5 with an average width of 1.19

- Average petal length of Setosa is much smaller than Versicolor and Virginica; however, the average sepal width of Setosa is greater than Versicolor and Virginica

- Petal length and width are strongly correlated, i.e., 96% of the time width increases with an increase in length.

- Petal length has a negative correlation with sepal width, i.e., 42% of the time increase in sepal width will decrease petal length.

- Initial conclusion from data: based on length and width of sepal/ petal alone, you can conclude that Versicolor/Virginica might resemble each other in size' however, Setosa characteristics seem to be noticeably different from the other two.

Further looking at the characteristics of the three Iris flowers visually in Figure 3-4, we can ascertain the hypothesis from our EDA.

setosa   versicolor   virginica

Figure 3-4. *Iris flowers*

Statistics and mathematics form the base for ML algorithms. Let's begin by understanding some of the basic concepts and algorithms that are derived from the statistical world and gradually move onto advanced ML algorithms.

Supervised Learning–Regression

Can you guess what is common in the set of business questions across different domains given in Table 3-6?

Table 3-6. *Supervised Learning Use Cases Examples*

Domain	Question
Retail	How much will be the daily, monthly, and yearly sales for a given store for the next 3 years?
Retail	How many car park spaces should be allocated to a retail store?
Manufacturing	How much will be the productwise manufacturing labor cost?
Manufacturing / Retail	How much will be my monthly electricity cost for the next 3 years?
Banking	What is the credit score of a customer?
Insurance	How many customers will claim the insurance this year?
Energy / Environmental	What will be the temperature for the next 5 days?

You might have guessed it right! The presence of the words 'how much' and 'how many' implies that the answer to these questions will be a quantitative or continuous number. Regression is one of the fundamental techniques that will help us to find the answer to these types of questions by studying the relationship between the different variables that are relevant to the questions.

Let's consider a use case where we have collected students' average test grade scores and their respective average studied hours for the test from a group of similar IQ students (Listing 3-9).

Listing 3-9. Students' Score vs. Hours Studied

```
import pandas as pd
import numpy as np
import matplotlib.pyplot as plt

# Load data
df = pd.read_csv('Data/Grade_Set_1.csv')
print(df)
```

```
# Simple scatter plot
df.plot(kind='scatter', x='Hours_Studied', y='Test_Grade', title='Grade vs
Hours Studied')
plt.show()
# check the correlation between variables
print("Correlation Matrix: ")
print(df.corr())
# ---- output ----
   Hours_Studied  Test_Grade
0              2          57
1              3          66
2              4          73
3              5          76
4              6          79
5              7          81
6              8          90
7              9          96
8             10         100
```

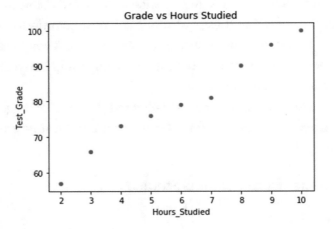

```
Correlation Matrix:
              Hours_Studied        Test_Grade
Hours_Studied     1.000000          0.987797
Test_Grade        0.987797          1.000000
```

A simple scatter plot with hours studied on the x-axis and the test grade on the y-axis shows that the grade gradually increases with an increase in hours studied. This implies that there is a linear relationship between the two variables. Further, performing the correlation analysis shows that there is a 98% positive relationship between the two variables; that means there's a 98% chance that any change in study hours will lead to a change in grade.

Correlation and Causation

Although correlation helps us determine the degree of relationship between two or more variables, it does not tell us about the cause and effect relationship. A high degree of correlation does not always necessarily mean a relationship of cause and effect exists between variables. Note that correlation does not imply causation, although the existence of causation always implies correlation. Let's understand this better with examples:

- More firemen's presence during a fire instance signifies that the fire is big, but the fire is not caused by firemen.

- When one sleeps with shoes on, one is likely to get a headache. This may be due to alcohol intoxication.

The significant degree of correlation in the preceding examples may be due to the following reasons:

- Small samples are prone to show a higher correlation due to pure chance.

- Variables may be influencing each other, so it becomes hard to designate one as the cause and the other the effect.

- Correlated variables may be influenced by one or more other related variables.

The domain knowledge or involvement of a subject matter expert is very important, to ascertain the correlation due to causation.

Fitting a Slope

Let's try to fit a slope line through all the points such that the error or residual (i.e., the distance of the line from each point) is best possible minimal (Figure 3-5).

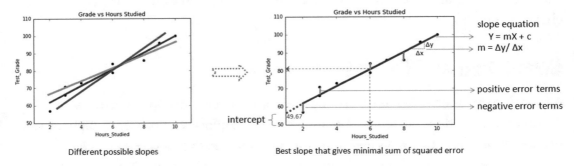

Figure 3-5. Linear Regression Model Components

The error could be positive or negative based on its location from the slope, because of which if we take a simple sum of all the errors it will be zero. So we should square the error to get rid of negativity and then sum the squared error. Hence, the slope is also referred to as the least squares line.

- The slope equation is given by Y = mX + c, where Y is the predicted value for a given x value.

- m is the change in y, divided by change in x (i.e., m is the slope of the line for the x variable and it indicates the steepness at which it increases with every unit increase in x variable value).

- c is the intercept, which indicates the location or point on the axis where it intersects; in the case of Figure 3-5 it is 49.67. The intercept is a constant that represents the variability in Y that is not explained by the X. It is the value of Y when X is zero.

Together the slope and intercept define the linear relationship between the two variables and can be used to predict or estimate an average rate of change. Now, using this relation for a new student, we can determine the score based on his/her study hours. Say a student is planning to study an overall 6 hours in preparation for the test. Simply drawing a connecting line from the x-axis and y-axis to the slope shows that there is a possibility of the student scoring 80. We can use the slope equation to predict the score for any given hours of study. In this case, the test grade is the dependent variable, denoted by "Y" and hours studied is the independent variable or predictor, denoted by "X." Let's use the linear regression function from the Scikit-learn library to find the values of m (x's coefficient) and c (intercept). Refer to Listing 3-10 for the example code.

Listing 3-10. Linear Regression

```
# Create linear regression object
lr = lm.LinearRegression()

x= df.Hours_Studied[:, np.newaxis] # independent variable
y= df.Test_Grade.values            # dependent variable

# Train the model using the training sets
lr.fit(x, y)
print("Intercept: ", lr.intercept_)
print("Coefficient: ", lr.coef_)

# manual prediction for a given value of x
print("Manual prediction :", 49.67777777777776 + 5.01666667*6)

# predict using the built-in function
print("Using predict function: ", lr.predict([[6]]))

# plotting fitted line
plt.scatter(x, y,  color='black')
plt.plot(x, lr.predict(x), color='blue', linewidth=3)
plt.title('Grade vs Hours Studied')
plt.ylabel('Test_Grade')
plt.xlabel('Hours_Studied')
# ---- output ----
Intercept:  49.67777777777776
Coefficient:  [5.01666667]
Manual prediction : 79.77777779777776
Using predict function:  [79.77777778]
```

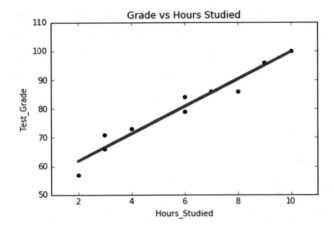

Let's put the appropriate values in the slope equation $(m * X + c = Y)$, $5.01 * 6 + 49.67 =$ 79.77; that means a student studying 6 hours has the probability of scoring a 79.77 test grade.

Note that if X is zero, the value of Y will be 49.67. That means even if the student does not study, there is a possibility that the score will be 49.67. This signifies that there are other variables that have a causation effect on the score that we do not have access to currently.

How Good Is Your Model?

There are three metrics widely used for evaluating linear model performance:

- R-Squared

- RMSE

- MAE

R-Squared for Goodness of fit

The R-Squared metric is the most popular practice of evaluating how well your model fits the data. R-Squared value designates the total proportion of variance in the dependent variable explained by the independent variable. It is a value between 0 and 1; the value toward 1 indicates a better model fit. Refer to Table 3-7 for R-Squared calculation illustration and Listing 3-11 for code implementation example.

Table 3-7. *Sample Table for R-Squared Calculation*

$$y \qquad \hat{y} \qquad (y_i - \bar{y})^2 \quad \sum (\hat{y}_i - \bar{y})^2$$

Hours_Studied	Test_Grade	Test_Grade_Pred	SST	SSR
2	57	59.71111	518.8272	402.6711
3	66	64.72778	189.8272	226.5025
4	73	69.74444	45.93827	100.6678
5	76	74.76111	14.2716	25.16694
6	79	79.77778	0.604938	0
7	81	84.79444	1.493827	25.16694
8	90	89.81111	104.4938	100.6678
9	96	94.82778	263.1605	226.5025
10	100	99.84444	408.9383	402.6711

Mean $(\bar{y})$ =   79.77

where

y dependent variable

$\hat{y}$ predicted variable

$\bar{y}$ mean of dependent variable

y_i i^{th} value of dependent variable column

$\hat{y}_i$ i^{th} value of predicted dependent variable column

```
                Total Sum of Square Residual (∑ SSR)
R-squared =  -----------------------------------------
                Sum of Square Total(∑ SST)

R-squared =  1510.01 / 1547.55 = 0.97
```

In this case, R-Squared can be interpreted as 97% of the variability in the dependent variable (test score) and can be explained by the independent variable (hours studied).

Root Mean Squared Error

This is the square root of the mean of the squared errors. RMSE indicates how close the predicted values are to the actual values; hence, lower RMSE value signifies that the model performance is good. One of the key properties of RMSE is that the unit will be the same as the target variable.

$$\sqrt{\frac{1}{n}\sum_{i=1}^{n}(y_i - \hat{y}_i)^2}$$

Mean Absolute Error

This (MAE) is the mean or average of the absolute value of the errors, that is, the predicted - actual.

$$\frac{1}{n}\sum_{i=1}^{n}|y_i - \hat{y}_i|$$

Listing 3-11. Linear Regression Model Accuracy Matrices

```
# function to calculate r-squared, MAE, RMSE
from sklearn.metrics import r2_score , mean_absolute_error,
mean_squared_error

# add predict value to the data frame
df['Test_Grade_Pred'] = lr.predict(x)

# Manually calculating R Squared
df['SST'] = np.square(df['Test_Grade'] - df['Test_Grade'].mean())
df['SSR'] = np.square(df['Test_Grade_Pred'] - df['Test_Grade'].mean())

print("Sum of SSR:", df['SSR'].sum())
print("Sum of SST:", df['SST'].sum())

print(df)
df.to_csv('r-squared.csv', index=False)

print("R Squared using manual calculation: ", df['SSR'].sum() / df['SST'].
sum()))
```

```
# Using built-in function
print("R Squared using built-in function: ", r2_score(df.Test_Grade,
df.Test_Grade_Pred))
print("Mean Absolute Error: ", mean_absolute_error(df.Test_Grade,
df.Test_Grade_Pred))
print("Root Mean Squared Error: ", np.sqrt(mean_squared_error(df.Test_
Grade, df.Test_Grade_Pred)))
# ---- output ----
Sum of SSR: 1510.01666667
Sum of SST: 1547.55555556
R Squared using manual calculation:  0.97574310741
R Squared using built-in function:  0.97574310741
Mean Absolute Error:  1.61851851852
Root Mean Squared Error:  2.04229959955
```

Outliers

Let's introduce an outlier): a student has studied 5 hours and scored 100. Assume that this student has a higher IQ than others in the group. Notice the drop in R-Squared value. So it is important to apply business logic to avoid including outliers in the training data set, to generalize the model and increase accuracy (Listing 3-12).

Listing 3-12. Outlier vs. R-Squared Value

```
# Load data
df = pd.read_csv('Data/Grade_Set_1.csv')

df.loc[9] = np.array([5, 100]) )

x= df.Hours_Studied[:, np.newaxis] # independent variable
y= df.Test_Grade.values          # dependent variable

# Train the model using the training sets
lr.fit(x, y)
print("Intercept: ", lr.intercept_)
print("Coefficient: ", lr.coef_)
```

```
# manual prediction for a given value of x
print("Manual prediction :", 54.4022988505747 + 4.64367816*6)

# predict using the built-in function
print("Using predict function: ", lr.predict([[6]]))

# plotting fitted line
plt.scatter(x, y,  color='black')
plt.plot(x, lr.predict(x), color='blue', linewidth=3)
plt.title('Grade vs Hours Studied')
plt.ylabel('Test_Grade')
plt.xlabel('Hours_Studied')

# add predict value to the data frame)
df['Test_Grade_Pred'] = lr.predict(x)

# Using built-in function
print("R Squared : ", r2_score(df.Test_Grade,  df.Test_Grade_Pred))
print("Mean Absolute Error: ", mean_absolute_error(df.Test_Grade,
df.Test_Grade_Pred))
print("Root Mean Squared Error: ", np.sqrt(mean_squared_error(df.Test_
Grade, df.Test_Grade_Pred)))
# ---- output ----
Intercept:  54.4022988505747
Coefficient:  [4.64367816]
Manual prediction : 82.2643678105747
Using predict function:  [82.26436782]
R Squared :  0.6855461390206965
Mean Absolute Error:  4.480459770114941)
Root Mean Squared Error:  7.761235830020588
```

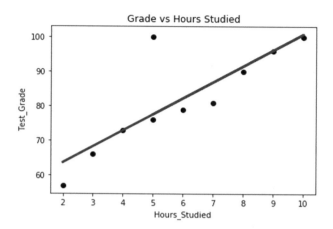

Polynomial Regression

It is a form of higher order linear regression modeled between dependent and independent variables as an nth degree polynomial. Although it's linear, it can fit curves better. Essentially we'll be introducing higher order degree variables of the same independent variable in the equation (Table 3-8 and Listing 3-13).

Table 3-8. *Polynomial Regression of Higher Degrees*

Degree	Regression Equation
Quadratic (2)	$Y = m_1X + m_2X^2 + c$
Cubic (3)	$Y = m_1X + m_2X^2 + m_3X^3 + c$
Nth	$Y = m_1X + m_2X^2 + m_3X^3 + \ldots m_nX^n + c$

Listing 3-13. Polynomial Regression

```
x = np.linspace(-3,3,1000) # 1000 sample number between -3 to 3

# Plot subplots
fig, ((ax1, ax2, ax3), (ax4, ax5, ax6)) = plt.subplots(nrows=2, ncols=3)

ax1.plot(x, x)
ax1.set_title('linear')
ax2.plot(x, x**2)
```

```
ax2.set_title('degree 2')
ax3.plot(x, x**3)
ax3.set_title('degree 3')
ax4.plot(x, x**4)
ax4.set_title('degree 4')
ax5.plot(x, x**5)
ax5.set_title('degree 5')
ax6.plot(x, x**6)
ax6.set_title('degree 6')

plt.tight_layout()# tidy layout
# --- output ----
```

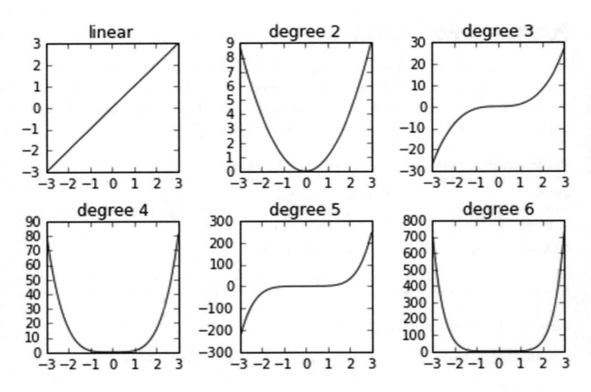

Let's consider another set of students' average test grade scores and their respective average studied hours for similar IQ students (Listing 3-14).

Listing 3-14. Polynomial Regression Example

```python
# importing linear regression function
import sklearn.linear_model as lm

# Load data
df = pd.read_csv('Data/Grade_Set_2.csv')
print(df)

# Simple scatter plot
df.plot(kind='scatter', x='Hours_Studied', y='Test_Grade', title='Grade vs
Hours Studied')

# check the correlation between variables
print("Correlation Matrix: ")
print(df.corr())

# Create linear regression object
lr = lm.LinearRegression()

x= df.Hours_Studied[:, np.newaxis]          # independent variable
y= df.Test_Grade                            # dependent variable

# Train the model using the training sets
lr.fit(x, y)

# plotting fitted line
plt.scatter(x, y,  color='black')
plt.plot(x, lr.predict(x), color='blue', linewidth=3)
plt.title('Grade vs Hours Studied')
plt.ylabel('Test_Grade')
plt.xlabel('Hours_Studied')

print("R Squared: ", r2_score(y, lr.predict(x)))
# ---- output ----
    Hours_Studied  Test_Grade
0             0.5          20
1             1.0          21
2             2.0          22
3             3.0          23
```

4	4.0	25
5	5.0	37
6	6.0	48
7	7.0	56
8	8.0	67
9	9.0	76
10	10.0	90
11	11.0	89
12	12.0	90

Correlation Matrix:

	Hours_Studied	Test_Grade
Hours_Studied	1.000000	0.974868
Test_Grade	0.974868	1.000000

R Squared: 0.9503677767

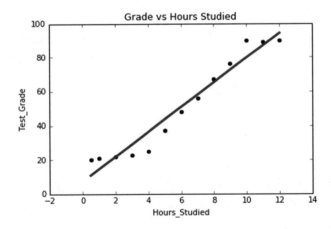

The correlation analysis shows a 97% positive relationship between hours studied and the test grade, and 95% (R-Squared) of variation in test grade can be explained by hours studied. Note that up to 4 hours of average study results in less than 30 test grade, and post 9 hours of study there is no value added to the grade. This is not a perfect linear relationship, although we can fit a linear line. Let's try higher order polynomial degrees (Listing 3-15).

Listing 3-15. R-Squared for Different Polynomial Degrees

```
lr = lm.LinearRegression()

x= df.Hours_Studied        # independent variable
y= df.Test_Grade           # dependent variable

# NumPy's vander function will return powers of the input vector
for deg in [1, 2, 3, 4, 5]:
    lr.fit(np.vander(x, deg + 1), y);
    y_lr = lr.predict(np.vander(x, deg + 1))
    plt.plot(x, y_lr, label='degree ' + str(deg));
    plt.legend(loc=2);
    print("R-squared for degree " + str(deg) + " = ",  r2_score(y, y_lr))
plt.plot(x, y, 'ok')
# ---- output ----
R-squared for degree 1 =  0.9503677767
R-squared for degree 2 =  0.960872656868
R-squared for degree 3 =  0.993832312037
R-squared for degree 4 =  0.99550001841
R-squared for degree 5 =  0.99562049139
```

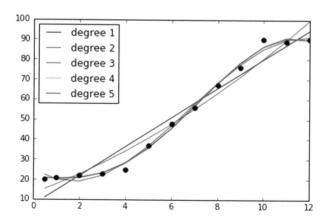

Note degree 1 here is the linear fit, and the higher order polynomial regression is fitting the curve better and R-Squared jumps 4% higher at degree 3. Beyond degree 3 there is not a massive change in R-Squared, so we can say that degree 3 fits better.

Scikit-learn provides a function to generate a new feature matrix consisting of all polynomial combinations of the features with a degree less than or equal to the specified degree (Listing 3-16).

Listing 3-16. Scikit-learn Polynomial Features

```
from sklearn.preprocessing import PolynomialFeatures
from sklearn.pipeline import make_pipeline

x= df.Hours_Studied[:, np.newaxis] # independent variable
y= df.Test_Grade                   # dependent variable

degree = 3
model = make_pipeline(PolynomialFeatures(degree), lr)

model.fit(x, y)

plt.scatter(x, y,  color='black')
plt.plot(x, model.predict(x), color='green')
plt.title('Grade vs Hours Studied')
plt.ylabel('Test_Grade')
plt.xlabel('Hours_Studied')

print("R Squared using built-in function: ", r2_score(y, model.predict(x)))
# ---- output ----
R Squared using built-in function:  0.993832312037
```

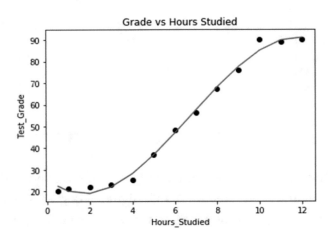

Multivariate Regression

So far we have seen simple regression with one independent variable for a given dependent variable. In most of the real-life use cases there will be more than one independent variable, so the concept of having multiple independent variables is called multivariate regression. The equation takes the following form.

$$y = m_1x_1 + m_2x_2 + m_3x_3 + ...+m_nx_n$$

where each independent variable is represented by xs, and ms are the corresponding coefficients. We'll be using the "statsmodels" Python library to learn the basics of multivariate regression because it provides more useful statistics results, which are helpful from a learning perspective. Once you understand the fundamental concepts, you can either use the Scikit-learn or statsmodels package, as both are efficient.

We'll be using the housing dataset (Table 3-9 from RDatasets), which contains sales prices of houses in the city of Windsor. Below is the brief description of each variable.

Table 3-9. *Housing Dataset (from RDatasets)*

Variable Name	Description	Data type
Price	The sale price of a house	Numeric
Lotsize	The lot size of property in square feet	Numeric
Bedrooms	Number of bedrooms	Numeric
Bathrms	Number of full bathrooms	Numeric
Stories	Number of stories excluding basement	Categorical
Driveway	Does the house have a driveway?	Boolean/Categorical
Recroom	Does the house have a recreational room?	Boolean/Categorical
Fullbase	Does the house have a full finished basement?	Boolean/Categorical
Gashw	Does the house use gas for hot water heating?	Boolean/Categorical
Airco	Does the house have central air conditioning?	Boolean/Categorical
Garagepl	Number of garage places	Numeric
Prefarea	Is the house located in the preferred neighborhood of the city?	Boolean/Categorical

Let's build a model to predict the house price (dependent variable) by considering the rest of the variables as independent variables.

The categorical variables need to be handled appropriately before running the first iteration of the model. Scikit-learn provides useful built-in preprocessing functions to handle categorical variables.

- *LabelBinarizer*: This will replace the binary variable text with numeric values. We'll be using this function for the binary categorical variables.

- *LabelEncoder*: This will replace category level with number representation.

- *OneHotEncoder*: This will convert n levels to an n-1 new variable, and the new variables will use 1 to indicate the presence of level and 0 for otherwise. Note that before calling OneHotEncoder, we should use LabelEncoder to convert levels to the number. Alternatively, we can achieve the same using the get_dummies of the Pandas package. This is more efficient to use, as we can directly use it on the column with text description without having to convert to numbers first.

Multicollinearity and Variation Inflation Factor

The dependent variable should have a strong relationship with independent variables. However, any independent variables should not have a strong correlation among other independent variables. Multicollinearity is an incident where one or more of the independent variables are strongly correlated with each other. In such an incident, we should use only one among correlated independent variables.

Multicollinearity and variance inflation factor (VIF) is an indicator of the existence of multicollinearity, and statsmodel provides a function to calculate the VIF for each independent variable' a value of greater than 10 is the rule of thumb for the possible existence of high multicollinearity. The standard guideline for VIF value is: VIF = 1 means no correlation exists, and VIF >1 but <5 means moderate correlation exists (Listing 3-17).

$$VIF_i = \frac{1}{1 - R_i^2}$$

where R_i^2 is the coefficient of determination of variable X_i.

Listing 3-17. Multicollinearity and VIF

```
# Load data
df = pd.read_csv('Data/Housing_Modified.csv')

# Convert binary fields to numeric boolean fields
lb = preprocessing.LabelBinarizer()

df.driveway = lb.fit_transform(df.driveway)
df.recroom = lb.fit_transform(df.recroom)
df.fullbase = lb.fit_transform(df.fullbase)
df.gashw = lb.fit_transform(df.gashw)
df.airco = lb.fit_transform(df.airco)
df.prefarea = lb.fit_transform(df.prefarea)

# Create dummy variables for stories
df_stories = pd.get_dummies(df['stories'], prefix='stories', drop_
first=True)

# Join the dummy variables to the main dataframe
df = pd.concat([df, df_stories], axis=1)
del df['stories']

# lets plot the correlation matrix using statmodels graphics packages'
plot_corr

# create correlation matrix
corr = df.corr()
sm.graphics.plot_corr(corr, xnames=list(corr.columns))
plt.show()
# ---- output ----
```

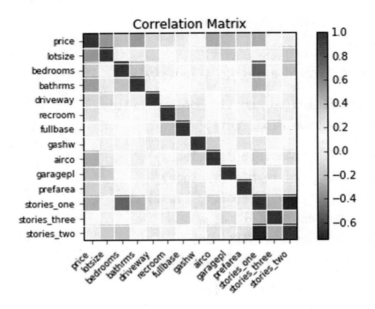

We can see from the plot that stories_one has a strong negative correlation with stories_two. Let's perform the VIF analysis to eliminate strongly correlated independent variables (Listing 3-18).

Listing 3-18. Remove Multicollinearity

```
from statsmodels.stats.outliers_influence import variance_inflation_factor,
OLSInfluence

# create a Python list of feature names
independent_variables = ['lotsize', 'bedrooms', 'bathrms','driveway', 'recroom',
                         'fullbase','gashw','airco','garagepl', 'prefarea',
                         'stories_one','stories_two','stories_three']

# use the list to select a subset from original DataFrame
X = df[independent_variables]
y = df['price']

thresh = 10

for i in np.arange(0,len(independent_variables)):
```

```
    vif = [variance_inflation_factor(X[independent_variables].values, ix)
for ix in range(X[independent_variables].shape[1])]
    maxloc = vif.index(max(vif))
    if max(vif) > thresh:
        print("vif :", vif)
        print('dropping \"' + X[independent_variables].columns[maxloc] + '\'
at index: ' + str(maxloc))
        del independent_variables[maxloc]
    else:
        break

print('Final variables:', independent_variables)
# ---- output ----
vif : [8.9580980878443359, 18.469878559519948, 8.9846723472908643,
7.0885785420918861, 1.4770152815033917, 2.013320236472385,
1.1034879198994192, 1.7567462065609021, 1.9826489313438442,
1.5332946465459893, 3.9657526747868612, 5.5117024083548918,
1.7700402770614867]
dropping 'bedrooms' at index: 1
Final variables: ['lotsize', 'bathrms', 'driveway', 'recroom', 'fullbase',
'gashw', 'airco', 'garagepl', 'prefarea', 'stories_one', 'stories_two',
'stories_three']
```

We can see that VIF analysis has eliminated bedrooms greater than 10; however, stories_one and stories_two have been retained.

Let's run the first iteration of a multivariate regression model with the set of independent variables that have passed the VIF analysis.

To test the model performance, the common practice is to split the data set into 80/20 (or 70/30) for train/test, respectively, and use the train data set to build the model. Then apply the trained model on the test data set to evaluate the performance of the model (Listing 3-19).

Listing 3-19. Build the Multivariate Linear Regression Model

```
from sklearn.model_selection import train_test_split
from sklearn import metrics
# create a Python list of feature names
```

```
independent_variables = ['lotsize', 'bathrms','driveway',
                         'fullbase','gashw', 'airco','garagepl',
                         'prefarea','stories_one','stories_three']

# use the list to select a subset from original DataFrame
X = df[independent_variables]

X_train, X_test, y_train, y_test = train_test_split(X, y, train_size=.80,
random_state=1)

# create a fitted model
lm = sm.OLS(y_train, X_train).fit()

# print the summary
print(lm.summary())

# make predictions on the testing set
y_train_pred = lm.predict(X_train)
y_test_pred = lm.predict(X_test)
y_pred = lm.predict(X) # full data
print("Train MAE: ", metrics.mean_absolute_error(y_train, y_train_pred))
print("Train RMSE: ", np.sqrt(metrics.mean_squared_error(y_train,
y_train_pred)))

print("Test MAE: ", metrics.mean_absolute_error(y_test, y_test_pred))
print("Test RMSE: ", np.sqrt(metrics.mean_squared_error(y_test,
y_test_pred)))
# ---- output ----
Train MAE:   11987.660160035877
Train RMSE:   15593.474917800835
Test MAE:   12722.079675396284
Test RMSE:   17509.25004003038
```

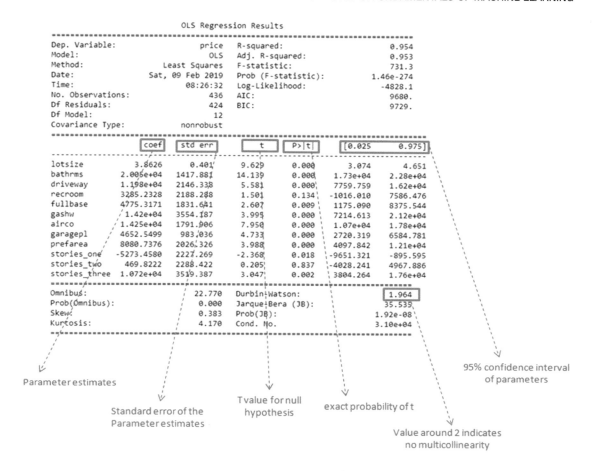

Interpreting the Ordinary Least Squares (OLS) Regression Results

Adjusted R-Squared: The simple R-Squared value will keep increasing with the addition of an independent variable. To fix this issue, adjusted R-Squared is considered for multivariate regression to understand the explanatory power of the independent variables.

$$Adjusted\,R^2 = 1 - \frac{(1-R^2)(N-1)}{N-p-1}$$

where N is total observations or sample size and p is the number of predictors.

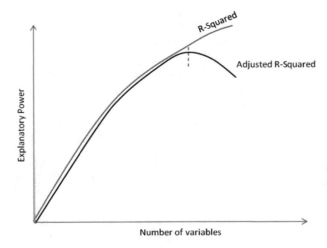

Figure 3-6. *R-Squared vs. Adjusted R-Squared*

- Figure 3-6 shows how R-Squared follows Adjusted R-Squared with increase of more variables

- With inclusion of more variables R-Squared always tend to increase

- Adjusted R-Squared will drop if the variable added does not explain the variable in the dependent variable

Coefficient: This is the individual coefficient for the respective independent variables. It can be either a positive or a negative number, which indicates that an increase in every unit of that independent variable will have a positive or negative impact on the dependent variable value.

Standard error: This is the average distance of the respective independent observed values from the regression line. The smaller values show that the model fitting is good.

Durbin-Watson: It's one of the common statistics used to determine the existence of multicollinearity, which means two or more independent variables used in the multivariate regression model are highly correlated. The Durbin-Watson statistic is always a number between 0 and 4. A value around 2 is ideal (range of 1.5 to 2.5 is relatively normal); it means that there is no autocorrelation between the variables used in the model.

Confidence interval: This is the coefficient to calculate a 95% confidence interval for the independent variable's slope.

t and p-value: p-value is one of the important statistics. In order to better understand, we'll have to explore the concept of hypothesis testing and normal distribution.

Hypothesis testing is an assertion regarding the distribution of the observations and validating this assertion. The hypothesis testing steps are as follows:

- A hypothesis is made.

- The validity of the hypothesis is tested.

- If the hypothesis is found to be true, it is accepted.

- If it is found to be untrue, it is rejected

- The hypothesis that is being tested for possible rejection is called the null hypothesis.

- The null hypothesis is denoted by H_0.

- The hypothesis that is accepted when the null hypothesis is rejected is called an alternate hypothesis H_a.

- The alternative hypothesis is often the interesting one and often the one that someone sets out to prove.

- For example, null hypothesis H0 is that the lot size has a real effect on house price; in this case, the coefficient m is equal to zero in the regression equation ($y = m * \text{lot size} + c$).

- Alternative hypothesis H_a is that the lot size does not have a real effect on house price, and the effect you saw was due to chance, which means the coefficient m is not equal to zero in the regression equation.

- In order to be able to say whether the regression estimate is close enough to the hypothesized value to be acceptable, we take the range of estimate implied by the estimated variance and see whether this range will contain the hypothesized value. To do this, we can transform the estimate into a standard normal distribution, and we know that 95% of all values of a variable that has a mean of 0 and variance of 1 will lie within 0 to 2 standard deviation. Given a regression estimate and its standard error, we can be 95% confident that the true (unknown) value of m will lie in this region (Figure 3-7).

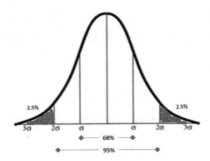

Figure 3-7. *Normal distribution (red is the rejection region)*

- The t-value is used to determine a p-value (probability), and p-value ≤0.05 signifies strong evidence against the null hypothesis, so you reject the null hypothesis. A p-value >0.05 signifies weak evidence against the null hypothesis, so you fail to reject the null hypothesis. So in our case, the variables with ≤0.05 means variables are significant for the model.

- The process of testing a hypothesis indicates that there is a possibility of making an error. There are two types of errors for any given data set, and these two types of errors are inversely related, which means the smaller the risk of one, the higher the risk of the other.

- *Type I error*: The error of rejecting the null hypothesis H0 even though H0 was true

- *Type II error*: The error of accepting the null hypothesis H0 even though H0 was false

- Note that variables "stories_three" and "recroom" have a large p-value, indicating it's insignificant. So let's rerun the regression without this variable and look at the results.

```
                           OLS Regression Results
==============================================================================
Dep. Variable:                    price   R-squared:                       0.954
Model:                              OLS   Adj. R-squared:                  0.953
Method:                   Least Squares   F-statistic:                     876.8
Date:                  Sat, 09 Feb 2019   Prob (F-statistic):           5.12e-277
Time:                          08:26:32   Log-Likelihood:                 -4829.2
No. Observations:                   436   AIC:                             9678.
Df Residuals:                       426   BIC:                             9719.
Df Model:                            10
Covariance Type:              nonrobust
==============================================================================
                 coef     std err          t      P>|t|      [0.025      0.975]
------------------------------------------------------------------------------
lotsize        3.9230       0.394      9.965      0.000       3.149       4.697
bathrms     2.017e+04    1302.611     15.482      0.000    1.76e+04    2.27e+04
driveway    1.224e+04    1992.869      6.141      0.000    8320.184    1.62e+04
fullbase    5729.3094    1691.457      3.387      0.001    2404.668    9053.951
gashw       1.432e+04    3542.864      4.043      0.000    7360.770    2.13e+04
airco       1.435e+04    1762.371      8.143      0.000    1.09e+04    1.78e+04
garagepl    4539.7003     965.076      4.704      0.000    2642.797    6436.603
prefarea    8261.1981    2021.190      4.087      0.000    4288.451    1.22e+04
stories_one -5762.8950   1549.027     -3.720      0.000   -8807.582   -2718.208
stories_three 1.03e+04   3101.467      3.320      0.001    4201.004    1.64e+04
==============================================================================
Omnibus:                         20.984   Durbin-Watson:                   1.962
Prob(Omnibus):                    0.000   Jarque-Bera (JB):               31.279
Skew:                             0.371   Prob(JB):                     1.61e-07
Kurtosis:                         4.082   Cond. No.                     2.67e+04
==============================================================================
```

Train MAE: 11993.3436816

Train RMSE: 15634.9995429

Test MAE: 12902.4799591

Test RMSE: 17694.9341405

- Note that dropping the variables has not impacted adjusted R-Squared negatively.

Regression Diagnostics

There are a set of procedures and assumptions that need to be verified about our model results, otherwise the model could be misleading. Let's look at some of the important regression diagnostics.

Outliers

Data points that are far away from the fitted regression line are called outliers, and these can impact the accuracy of the model. Plotting normalized residual vs. leverage will give us a good understanding of the outlier's points. Residual is the difference between actual vs. predicted, and leverage is a measure of how far away the independent variable values of observation are from those of the other observations (Listing 3-20).

Listing 3-20. Plot the Normalized Residual vs. Leverage

```
# lets plot the normalized residual vs leverage
from statsmodels.graphics.regressionplots import plot_leverage_resid2
fig, ax = plt.subplots(figsize=(8,6))
fig = plot_leverage_resid2(lm, ax = ax)
# ---- output ----
```

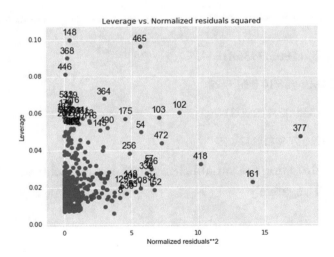

From the chart, we see that there are many observations that have high leverage and residual. Running a Bonferroni outlier test will give us p-values for each observation, and those observations with p-value <0.05 are the outliers affecting the accuracy. It is a good practice to consult or apply business domain knowledge to make a decision on removing the outlier points and rerunning the model; these points could be natural in the process, although they are mathematically found as outliers (Listing 3-21).

Listing 3-21. Find Outliers

```
# Find outliers #
# Bonferroni outlier test
test = lm.outlier_test()

print('Bad data points (bonf(p) < 0.05):')
print(test[test['bonf(p)'] < 0.05])
# ---- output ----
Bad data points (bonf(p) < 0.05):
     student_resid    unadj_p    bonf(p)
377        4.387449   0.000014   0.006315
```

Homoscedasticity and Normality

The error variance should be constant, which is known as homoscedasticity, and the error should be normally distributed (Listing 3-22).

Listing 3-22. Homoscedasticity Test

```
# plot to check homoscedasticity
plt.plot(lm.resid,'o')
plt.title('Residual Plot')
plt.ylabel('Residual')
plt.xlabel('Observation Numbers')
plt.show()
plt.hist(lm.resid, normed=True)
# ---- output ----
```

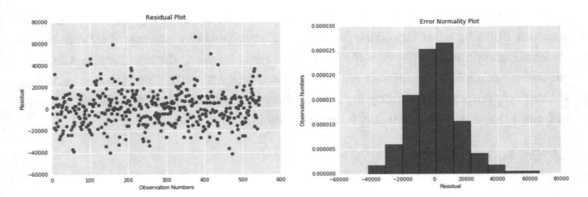

The relationships between the predictors and the outcome variable should be linear. If the relationship is not linear then appropriate transformation (such as log, square root, and higher order polynomials, etc.) should be applied to the dependent/independent variable to fix the issue (Listing 3-23).

Listing 3-23. Linearity Check

```
# linearity plots
fig = plt.figure(figsize=(10,15))
fig = sm.graphics.plot_partregress_grid(lm, fig=fig)
# ---- output ----
```

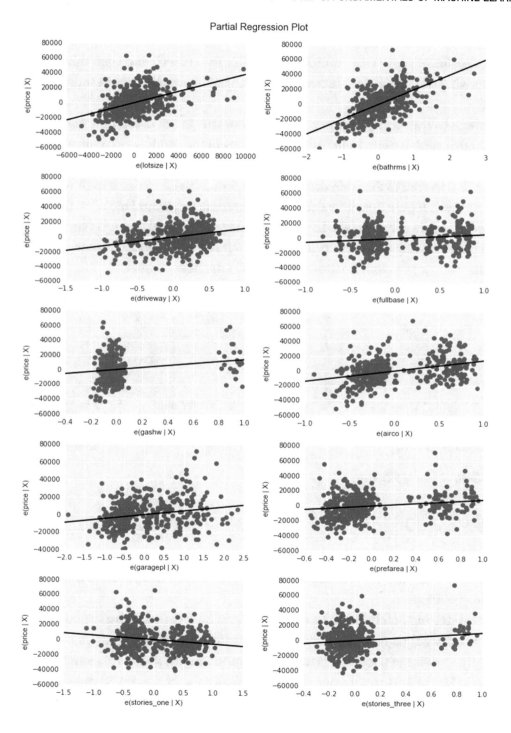

Partial Regression Plot

Overfitting and Underfitting

Underfitting occurs when the model does not fit the data well and is unable to capture the underlying trend in it. In this case, we can notice a low accuracy in training and test data set.

Conversely, overfitting occurs when the model fits the data too well, capturing all the noises. In this case, we can notice a high accuracy in the training data set, whereas the same model will result in a low accuracy on the test data set. This means the model has fitted the line so well to the train data set that it failed to generalize it to fit well on the unseen data set. Figure 3-8 shows how the different fitting would look like for an earlier discussed example use case. Choice of right order polynomial degree is very important to avoid an overfitting or underfitting issue in regression. We'll also discuss in detail about different ways to handle these problems in the next chapter.

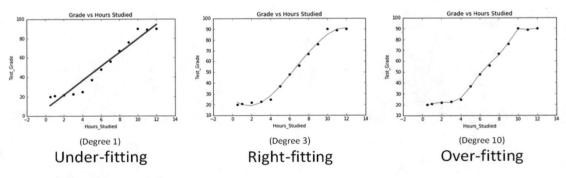

Figure 3-8. *Model fittings*

Regularization

With the increase in a number of variables and the increase in model complexity, the probability of overfitting also increases. Regularization is a technique to avoid the overfitting problem.

Statsmodel and Scikit-learn provide Ridge and LASSO (least absolute shrinkage and selection operator) regression to handle the overfitting issue. With the increase in model complexity, the size of coefficients increases exponentially, so the ridge and LASSO regression (Figure 3-9) apply a penalty to the magnitude of the coefficient to handle the issue. Refer to Listing 3-24 for example code implementation.

LASSO: This provides a sparse solution, also known as L1 regularization. It guides parameter value to be zero (i.e., the coefficients of the variables that add minor value to the model will be zero), and adds a penalty equivalent to the absolute value of the magnitude of coefficients.

Ridge regression: Also known as Tikhonov (L2) regularization, it guides parameters to be close to zero but not zero. You can use this when you have many variables that add minor value to the model accuracy individually, however overall improve the model accuracy and cannot be excluded from the model. Ridge regression will apply a penalty to reduce the magnitude of the coefficient of all variables that add minor value to the model accuracy, adding penalty equivalent to the square of the magnitude of coefficients. Alpha is the regularization strength and must be a positive float.

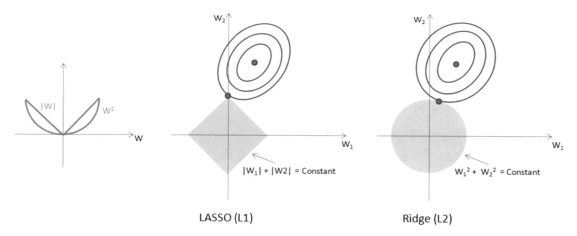

Figure 3-9. Regularizations

Listing 3-24. Regularization

```
from sklearn import linear_model

# Load data
df = pd.read_csv('Data/Grade_Set_2.csv')
df.columns = ['x','y']
```

```python
for i in range(2,50):                    # power of 1 is already there
    colname = 'x_%d'%i                   # new var will be x_power
    df[colname] = df['x']**i

independent_variables = list(df.columns)
independent_variables.remove('y')

X= df[independent_variables]        # independent variable
y= df.y                              # dependent variable

# split data into train and test
X_train, X_test, y_train, y_test = train_test_split(X, y, train_size=.80,
random_state=1)

# Ridge regression
lr = linear_model.Ridge(alpha=0.001)
lr.fit(X_train, y_train)
y_train_pred = lr.predict(X_train)
y_test_pred = lr.predict(X_test)

print("------ Ridge Regression ------")
print("Train MAE: ", metrics.mean_absolute_error(y_train, y_train_pred))
print("Train RMSE: ", np.sqrt(metrics.mean_squared_error(y_train,
y_train_pred)))

print("Test MAE: ", metrics.mean_absolute_error(y_test, y_test_pred))
print("Test RMSE: ", np.sqrt(metrics.mean_squared_error(y_test,
y_test_pred)))
print("Ridge Coef: ", lr.coef_)

# LASSO regression
lr = linear_model.Lasso(alpha=0.001)
lr.fit(X_train, y_train)
y_train_pred = lr.predict(X_train)
y_test_pred = lr.predict(X_test)

print("----- LASSO Regression -----")
print("Train MAE: ", metrics.mean_absolute_error(y_train, y_train_pred))
```

```
print("Train RMSE: ", np.sqrt(metrics.mean_squared_error(y_train,
y_train_pred)))

print("Test MAE: ", metrics.mean_absolute_error(y_test, y_test_pred))
print("Test RMSE: ", np.sqrt(metrics.mean_squared_error(y_test,
y_test_pred)))
print("LASSO Coef: ", lr.coef_)
#--- output ----
------ Ridge Regression ------
Train MAE:   12.775326528414379
Train RMSE:   16.72063936357992
Test MAE:   22.397943556789926
Test RMSE:   22.432642089791898
Ridge Coef:  [ 1.01446487e-88  1.27690319e-87  1.41113660e-86  1.49319913e-85
   1.54589299e-84  1.58049535e-83  1.60336716e-82  1.61825366e-81
   1.62742313e-80  1.63228352e-79  1.63372709e-78  1.63232721e-77
   1.62845333e-76  1.62233965e-75  1.61412730e-74  1.60389073e-73
   1.59165478e-72  1.57740595e-71  1.56110004e-70  1.54266755e-69
   1.52201757e-68  1.49904080e-67  1.47361205e-66  1.44559243e-65
   1.41483164e-64  1.38117029e-63  1.34444272e-62  1.30448024e-61
   1.26111524e-60  1.21418622e-59  1.16354417e-58  1.10906042e-57
   1.05063662e-56  9.88217010e-56  9.21803842e-55  8.51476330e-54
   7.77414158e-53  6.99926407e-52  6.19487106e-51  5.36778815e-50
   4.52745955e-49  3.68659929e-48  2.86198522e-47  2.07542549e-46
   1.35493365e-45  7.36155358e-45  2.64098894e-44 -4.76790286e-45
   2.09597530e-46]
----- LASSO Regression -----
Train MAE:   0.8423742988874519
Train RMSE:   1.219129185560593
Test MAE:   4.32364759404346
Test RMSE:   4.872324349696696
LASSO Coef:  [ 1.29948409e+00  3.92103580e-01  1.75369422e-02  7.79647589e-04
   3.02339084e-05  3.35699852e-07 -1.13749601e-07 -1.79773817e-08
  -1.93826156e-09 -1.78643532e-10 -1.50240566e-11 -1.18610891e-12
  -8.91794276e-14 -6.43309631e-15 -4.46487394e-16 -2.97784537e-17
  -1.89686955e-18 -1.13767046e-19 -6.22157254e-21 -2.84658206e-22
```

```
 -7.32019963e-24   5.16015995e-25   1.18616856e-25   1.48398312e-26
  1.55203577e-27   1.48667153e-28   1.35117812e-29   1.18576052e-30
  1.01487234e-31   8.52473862e-33   7.05722034e-34   5.77507464e-35
  4.68162529e-36   3.76585569e-37   3.00961249e-38   2.39206785e-39
  1.89235649e-40   1.49102460e-41   1.17072537e-42   9.16453614e-44
  7.15512017e-45   5.57333358e-46   4.33236496e-47   3.36163309e-48
  2.60423554e-49   2.01461728e-50   1.55652093e-51   1.20123190e-52
  9.26105400e-54]
```

Nonlinear Regression

Linear models are mostly linear in nature, although they need not be straight fitting. In contrast, the nonlinear model's fitted line can take any shape' this scenario usually occurs when models are derived on the basis of physical or biological considerations. The nonlinear model has a direct interpretation in terms of the process under study. The SciPy library provides a curve_fit function to fit models to scientific data based on theory, to determine the parameters of a physical system. Some of the example use cases are Michaelis–Menten's enzyme kinetics, Weibull distribution, and power law distribution (Listing 3-25).

Listing 3-25. Nonlinear Regression

```python
import numpy as np
import matplotlib.pyplot as plt
from scipy.optimize import curve_fit
%matplotlib inline

y = np.array([1.0, 1.5, 2.4, 2, 1.49, 1.2, 1.3, 1.2, 0.5])

# Function for non-liear curve fitting
def func(x, p1,p2):
    return p1*np.sin(p2*x) + p2*np.cos(p1*x)

popt, pcov = curve_fit(func, x, y,p0=(1.0,0.2))

p1 = popt[0]
p2 = popt[1]
residuals = y - func(x,p1,p2)
fres = sum(residuals**2)
```

```
curvex=np.linspace(-2,3,100)
curvey=func(curvex,p1,p2)

plt.plot(x,y,'bo ')
plt.plot(curvex,curvey,'r')
plt.title('Non-linear fitting')
plt.xlabel('x')
plt.ylabel('y')
plt.legend(['data','fit'],loc='best')
plt.show()
# ---- output ----
```

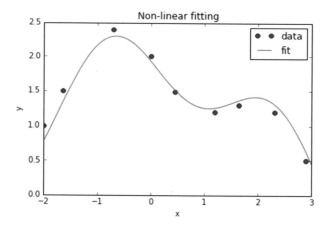

Supervised Learning–Classification

Let's look at another set of questions (Table 3-10). Can you guess what is common in this set of business questions across different domains?

Table 3-10. *Classification Use Case Examples*

Domain	Question
Telecom	Is a customer likely to leave the network? (churn prediction)
Retail	Is he a prospective customer (i.e., the likelihood of purchases vs. nonpurchase)?
Insurance	To issue insurance, should a customer be sent for a medical checkup?
Insurance	Will the customer renew the insurance?
Banking	Will a customer default the loan amount?
Banking	Should a customer be given a loan?
Manufacturing	Will the equipment fail?
Healthcare	Is the patient infected with a disease?
Healthcare	What type of disease does a patient have?
Entertainment	What is the genre of music?

The answers to these questions are a discrete class. The number of level or class can vary from a minimum of two (example: true or false, yes or no) to multiclass. In ML, classification deals with identifying the probability a new object is a member of a class or set. The classifiers are the algorithms that map the input data (also called features) to categories.

Logistic Regression

Let's consider a use case where we have to predict students test outcome: pass (1) or fail (0) based on hours studied. In this case, the outcome to be predicted is discrete. Let's build a linear regression and try to use a threshold: anything over some value is a pass, else fail (Listing 3-26).

Listing 3-26. Logistic Regression

```
import sklearn.linear_model as lm

# Load data
df = pd.read_csv('Data/Grade_Set_1_Classification.csv')
```

```
print (df)
x= df.Hours_Studied[:, np.newaxis] # independent variable
y= df.Result                       # dependent variable

# Create linear regression object
lr = lm.LinearRegression()

# Train the model using the training sets
lr.fit(x, y)

# plotting fitted line
plt.scatter(x, y,  color='black')
plt.plot(x, lr.predict(x), color='blue', linewidth=3)
plt.title('Hours Studied vs Result')
plt.ylabel('Result')
plt.xlabel('Hours_Studied')

# add predict value to the data frame
df['Result_Pred'] = lr.predict(x)

# Using built-in function
print ("R Squared : ", r2_score(df.Result, df.Result_Pred))
print ("Mean Absolute Error: ", mean_absolute_error(df.Result, df.Result_
Pred))
print ("Root Mean Squared Error: ", np.sqrt(mean_squared_error(df.Result,
df.Result_Pred)))
# ---- output ----
R Squared :  0.675
Mean Absolute Error:  0.22962962963
Root Mean Squared Error:  0.268741924943
```

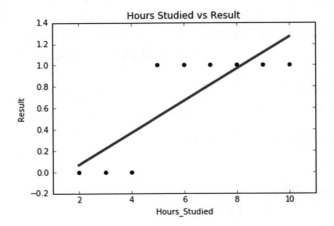

The outcome that we are expecting is either 1 or 0; the issue with linear regression is that it can give values larger than 1 or less than 0. In the preceding plot, we can see that linear regression is not able to draw boundaries to classify observations.

The solution to this is to introduce a sigmoid or logit function (which takes an S shape) to the regression equation. The fundamental idea here is that the hypothesis will use the linear approximation, then map with a logistic function for binary prediction. The linear regression equation in this case is y = mx + c.

Logistic regression can be explained better in the odds ratio. The odds of an event occurring are defined as the probability of an event occurring divided by the probability of that event not occurring.

$$\text{odds ratio of pass vs fail} = probability(y = 1) \, / \, 1 - probability(y = 1)$$

A logit is the log base e(log) of the odds, so using a logit model:

$$\log(p \, / \, p(1 - p)) = mx + c$$

Figure 3-10 shows the logistic regression equation probability $(y = 1) = 1 \, / \, 1 + e^{-(mx + c)}$ and Listing 3-27 shows the code implementation to plot the sigmoid.

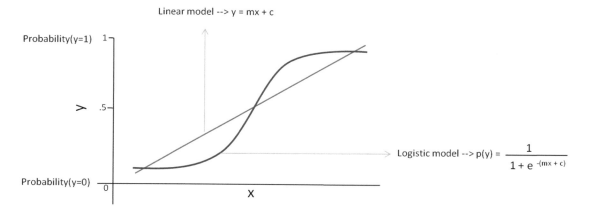

Figure 3-10. *Linear regression vs. logistic regression*

Listing 3-27. Plot Sigmoid Function

```
# plot sigmoid function
x = np.linspace(-10, 10, 100)
y = 1.0 / (1.0 + np.exp(-x))

plt.plot(x, y, 'r-', label='logit')
plt.legend(loc='lower right')
# --- output ----
```

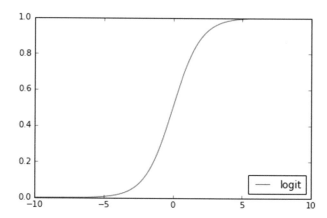

Listing 3-28 shows an example code implementation for logistic regression using the Scikit-learn package.

Listing 3-28. Logistic Regression Using Scikit-learn

```
from sklearn.linear_model import LogisticRegression

# manually add intercept
df['intercept'] = 1
independent_variables = ['Hours_Studied', 'intercept']

x = df[independent_variables]        # independent variable
y = df['Result']                     # dependent variable

# instantiate a logistic regression model, and fit with X and y
model = LogisticRegression()
model = model.fit(x, y)

# check the accuracy on the training set
model.score(x, y)

# predict_proba will return array containing probability of y = 0 and y = 1
print ('Predicted probability:', model.predict_proba(x)[:,1])

# predict will give convert the probability(y=1) values > .5 to 1 else 0
print ('Predicted Class:',model.predict(x))

# plotting fitted line
plt.scatter(df.Hours_Studied, y,  color='black')
plt.yticks([0.0, 0.5, 1.0])
plt.plot(df.Hours_Studied, model.predict_proba(x)[:,1], color='blue',
linewidth=3)
plt.title('Hours Studied vs Result')
plt.ylabel('Result')
plt.xlabel('Hours_Studied')
plt.show()
# ---- output ----
Predicted probability: [0.38623098 0.49994056 0.61365629 0.71619252
0.80036836 0.86430823 0.91006991 0.94144416 0.96232587]
Predicted Class: [0 0 1 1 1 1 1 1 1]
```

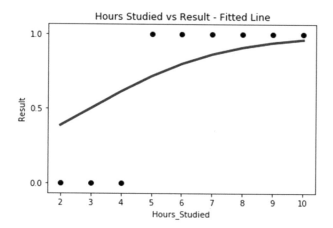

Evaluating a Classification Model Performance

The confusion matrix is the table that is used for describing the performance of the classification model. Figure 3-11 shows the confusion matrix.

		Predicted	
		FALSE	TRUE
Actual	FALSE	TN = 2	FP = 1
	TRUE	FN = 0	TP = 6

Figure 3-11. *Confusion matrix*

- True Negative (TN): Actual FALSE that was predicted as FALSE

- False Positive (FP): Actual FALSE that was predicted as TRUE (Type I error)

- False Negative (FN): Actual TRUE that was predicted as FALSE (Type II error)

- True Positive (TP): Actual TRUE that was predicted as TRUE

Ideally, a good model should have high TN and TP and less of Type I & II errors. Table 3-11 describes the key metrics derived out of a confusion matrix to measure the classification model performance. (Table 3-11). Listing 3-29 is the code example to produce a confusion matrix.

Table 3-11. *Classification Pefromance Matrices*

Metric	Description	Formula
Accuracy	What % of predictions was correct?	(TP+TN)/(TP+TN+FP+FN)
Misclassification Rate	What % of prediction is wrong?	(FP+FN)/(TP+TN+FP+FN)
True Positive Rate OR SensitivityORRecall (completeness)	What % of positive cases did model catch?	TP/(FN+TP)
False Positive Rate	What % of No was predicted as Yes?	FP/(FP+TN)
Specificity	What % of No was predicted as No?	TN/(TN+FP)
Precision (exactness)	What % of positive predictions was correct?	TP/(TP+FP)
F1 score	Weighted average of precision and recall	2∗((precision ∗ recall) / (precision + recall))

Listing 3-29. Confusion Matrix

```
from sklearn import metrics

# generate evaluation metrics
print ("Accuracy :", metrics.accuracy_score(y, model.predict(x)))
print ("AUC :", metrics.roc_auc_score(y, model.predict_proba(x)[:,1]))

print ("Confusion matrix :",metrics.confusion_matrix(y, model.predict(x)))
print ("classification report :", metrics.classification_report(y, model.
predict(x)))
# ----output----
Accuracy : 0.88
AUC : 1.0
Confusion matrix : [[2 1] [0 6]]
```

```
classification report :
               precision    recall   f1-score    support

           0       1.00       0.67      0.80         3
           1       0.86       1.00      0.92         6


   micro avg        0.89       0.89      0.89         9
   macro avg        0.93       0.83      0.86         9
weighted avg        0.90       0.89      0.88         9
```

ROC Curve

An ROC (receiver operating characteristic)curve is one more important metric, a most commonly used way to visualize the performance of a binary classifier; and AUC is believed to be one of the best ways to summarize performance in a single number. AUC indicates that the probability of a randomly selected positive example will be scored higher by the classifier than a randomly selected negative example. If you have multiple models with nearly the same accuracy, you can pick the one that gives higher AUC (Listing 3-30).

Listing 3-30. Area Under the Curve

```
# Determine the false positive and true positive rates
fpr, tpr, _ = metrics.roc_curve(y, model.predict_proba(x)[:,1])

# Calculate the AUC
roc_auc = metrics.auc(fpr, tpr)
print ('ROC AUC: %0.2f' % roc_auc)

# Plot of a ROC curve for a specific class
plt.figure()
plt.plot(fpr, tpr, label='ROC curve (area = %0.2f)' % roc_auc)
plt.plot([0, 1], [0, 1], 'k--')
plt.xlim([0.0, 1.0])
plt.ylim([0.0, 1.05])
plt.xlabel('False Positive Rate')
plt.ylabel('True Positive Rate')
```

```
plt.title('ROC Curve')
plt.legend(loc="lower right")
plt.show()
#---- output ----
```

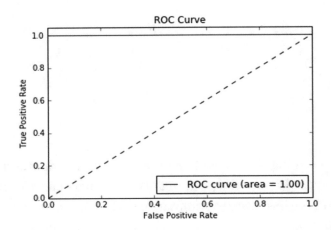

In the preceding case AUC is 100%, as the model is able to predict all the positive instances as a true positive.

Fitting Line

The inverse of regularization is one of the key aspects of fitting a logistic regression line. It defines the complexity of the fitted line. Let's try to fit lines for different values for this parameter (C, default is 1) and see how the fitting line and the accuracy change (Listing 3-31).

Listing 3-31. Controling Complexity for Fitting a Line

```
# instantiate a logistic regression model with default c value, and fit
with X and y
model = LogisticRegression()
model = model.fit(x, y)

# check the accuracy on the training set
print ("C = 1 (default), Accuracy :", metrics.accuracy_score(y, model.
predict(x)))
```

```
# instantiate a logistic regression model with c = 10, and fit with X and y
model1 = LogisticRegression(C=10)
model1 = model1.fit(x, y)

# check the accuracy on the training set
print ("C = 10, Accuracy :", metrics.accuracy_score(y, model1.predict(x)))

# instantiate a logistic regression model with c = 100, and fit with X and y
model2 = LogisticRegression(C=100)
model2 = model2.fit(x, y)

# check the accuracy on the training set
print ("C = 100, Accuracy :", metrics.accuracy_score(y, model2.predict(x)))

# instantiate a logistic regression model with c = 1000, and fit with X and y
model3 = LogisticRegression(C=1000)
model3 = model3.fit(x, y)

# check the accuracy on the training set
print ("C = 1000, Accuracy :", metrics.accuracy_score(y, model3.predict(x)))

# plotting fitted line
plt.scatter(df.Hours_Studied, y,  color='black', label='Result')
plt.yticks([0.0, 0.5, 1.0])
plt.plot(df.Hours_Studied, model.predict_proba(x)[:,1], color='gray',
linewidth=2, label='C=1.0')
plt.plot(df.Hours_Studied, model1.predict_proba(x)[:,1], color='blue',
linewidth=2,label='C=10')
plt.plot(df.Hours_Studied, model2.predict_proba(x)[:,1], color='green',
linewidth=2,label='C=100')
plt.plot(df.Hours_Studied, model3.predict_proba(x)[:,1], color='red',
linewidth=2,label='C=1000')
plt.legend(loc='lower right') # legend location
plt.title('Hours Studied vs Result')
plt.ylabel('Result')
plt.xlabel('Hours_Studied')
plt.show()
```

```
#----output----
C = 1 (default), Accuracy : 0.88
C = 10, Accuracy : 1.0
C = 100, Accuracy : 1.0
C = 1000, Accuracy : 1.0
```

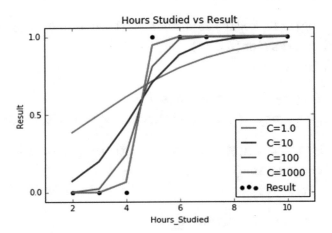

Stochastic Gradient Descent

Fitting the right slope that minimizes the error (also known as cost function) for a large data set can be tricky. However, this can be achieved through a stochastic gradient descent (steepest descent) optimization algorithm. In the case of regression problems, the cost function J to learn the weights can be defined as the sum of squared errors (SSE) between actual vs. predicted value.

$$J(w) = \frac{1}{2}\sum_i (y^i - \hat{y}^i),$$ where y^i the ith is an actual value, and $\hat{y}^i$ is the ith predicted value.

The stochastic gradient descent algorithm to update weight (w), for every weight j of every training sample I, can be given as (repeat until convergence)

$$\left\{W_j := W_j + \alpha \sum_{i=1}^{m}(y^i - \hat{y}^i)x_j^i\right\}.$$ Alpha (α) is the learning rate, and choosing a smaller value

for the same will ensure that the algorithm does not miss global cost minimum (Figure 3-12).

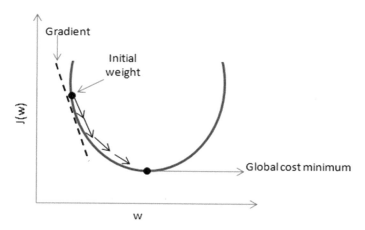

Figure 3-12. *Gradient descent*

The default solver parameter for logistic regression in Scikit-learn is "liblinear." which works fine for the smaller dataset. For a large dataset with a large number of independent variables, "sag" (stochastic average gradient descent) is the recommended solver to fit the optimal slope faster.

Regularization

With an increase in the number of variables, the probability of overfitting also increases. LASSO (L1) and Ridge (L2) can be applied for logistic regression as well, to avoid overfitting. Let's look at an example to understand the over/underfitting issue in logistic regression (Listing 3-32).

Listing 3-32. Underfitting, Right-Fitting, and Overfitting

```
import pandas as pd
data = pd.read_csv('Data\LR_NonLinear.csv')

pos = data['class'] == 1
neg = data['class'] == 0
x1 = data['x1']
x2 = data['x2']

# function to draw scatter plot between two variables
def draw_plot():
    plt.figure(figsize=(6, 6))
```

```
    plt.scatter(np.extract(pos, x1),
                np.extract(pos, x2),
                c='b', marker='s', label='pos')
    plt.scatter(np.extract(neg, x1),
                np.extract(neg, x2),
                c='r', marker='o', label='neg')
    plt.xlabel('x1');
    plt.ylabel('x2');
    plt.axes().set_aspect('equal', 'datalim')
    plt.legend();

# create hihger order polynomial for independent variables
order_no = 6

# map the variable 1 & 2 to its higher order polynomial
def map_features(variable_1, variable_2, order=order_no):
    assert order >= 1
    def iter():
        for i in range(1, order + 1):
            for j in range(i + 1):
                yield np.power(variable_1, i - j) * np.power(variable_2, j)
    return np.vstack(iter())

out = map_features(data['x1'], data['x2'], order=order_no)
X = out.transpose()
y = data['class']

# split the data into train and test
X_train, X_test, y_train, y_test = train_test_split(X, y, test_size=0.3,
random_state=0)

# function to draw classifier line
def draw_boundary(classifier):
    dim = np.linspace(-0.8, 1.1, 100)
    dx, dy = np.meshgrid(dim, dim)
    v = map_features(dx.flatten(), dy.flatten(), order=order_no)
    z = (np.dot(classifier.coef_, v) + classifier.intercept_).reshape(100, 100)
    plt.contour(dx, dy, z, levels=[0], colors=['r'])
```

```
# fit with c = 0.01
clf = LogisticRegression(C=0.01).fit(X_train, y_train)
print ('Train Accuracy for C=0.01: ', clf.score(X_train, y_train))
print ('Test Accuracy for C=0.01: ', clf.score(X_test, y_test))
draw_plot()
plt.title('Fitting with C=0.01')
draw_boundary(clf)
plt.legend();

# fit with c = 1
clf = LogisticRegression(C=1).fit(X_train, y_train)
print ('Train Accuracy for C=1: ', clf.score(X_train, y_train))
print ('Test Accuracy for C=1: ', clf.score(X_test, y_test))
draw_plot()
plt.title('Fitting with C=1')
draw_boundary(clf)
plt.legend();

# fit with c = 10000
clf = LogisticRegression(C=10000).fit(X_train, y_train)
print ('Train Accuracy for C=10000: ', clf.score(X_train, y_train))
print ('Test Accuracy for C=10000: ', clf.score(X_test, y_test))
draw_plot()
plt.title('Fitting with C=10000')
draw_boundary(clf)
plt.legend();
#----output----
Train Accuracy for C=0.01:  0.624242424242
Test Accuracy for C=0.01:  0.619718309859
Train Accuracy for C=1:  0.842424242424
Test Accuracy for C=1:  0.859154929577
Train Accuracy for C=10000:  0.860606060606
Test Accuracy for C=10000:  0.788732394366
```

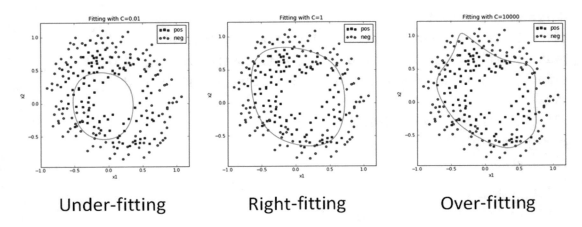

Under-fitting   Right-fitting   Over-fitting

Notice that with higher order regularization, value overfitting occurs. The same can be determined by looking at the accuracy between train and test datasets (i.e., accuracy drops significantly in the test data set).

Multiclass Logistic Regression

Logistic regression can also be used to predict the dependent or target variable with multiclass. Let's learn multiclass prediction with the Iris data set, one of the best-known databases to be found in the pattern recognition literature. The data set contains three classes of 50 instances each, where each class refers to a type of Iris plant. This comes as part of the Scikit-learn datasets, where the third column represents the petal length, and the fourth column represents petal width of the flower samples. The classes are already converted to integer labels, where 0 = Iris-Setosa, 1 = Iris-Versicolor, 2 = Iris-Virginica.

Load Data

We can load the data from sklearn datasets as shown in Listing 3-33.

Listing 3-33. Load Data

```
from sklearn import datasets
import numpy as np
import pandas as pd
iris = datasets.load_iris()
X = iris.data
y = iris.target
```

```
print('Class labels:', np.unique(y))
#----output----
('Class labels:', array([0, 1, 2]))
```

Normalize Data

The unit of measurement might differ, so let's normalize the data before building the model (Listing 3-34).

Listing 3-34. Normalize Data

```
from sklearn.preprocessing import StandardScaler
sc = StandardScaler()
sc.fit(X)
X = sc.transform(X)
```

Split Data

Split data into train and test. Whenever we are using a random function, it's advised to use a seed to ensure the reproducibility of the results (Listing 3-35).

Listing 3-35. Split Data into Train and Test

```
# split data into train and test
from sklearn.model_selection import train_test_split
X_train, X_test, y_train, y_test = train_test_split(X, y, test_size=0.3,
random_state=0)
```

Training Logistic Regression Model and Evaluating

Listing 3-36 is the example code implementation for logistic regression model training and evaluation.

Listing 3-36. Logistic Regression Model Training and Evaluation

```
from sklearn.linear_model import LogisticRegression

# l1 regularization gives better results
lr = LogisticRegression(penalty='l1', C=10, random_state=0)
lr.fit(X_train, y_train)

from sklearn import metrics

# generate evaluation metrics
print("Train - Accuracy :", metrics.accuracy_score(y_train, lr.predict
(X_train)))
print("Train - Confusion matrix :",metrics.confusion_matrix(y_train,
lr.predict(X_train)))
print("Train - classification report :", metrics.classification_report
(y_train, lr.predict(X_train)))

print("Test - Accuracy :", metrics.accuracy_score(y_test, lr.predict
(X_test)))
print("Test - Confusion matrix :",metrics.confusion_matrix(y_test,
lr.predict(X_test)))
print("Test - classification report :", metrics.classification_report
(y_test, lr.predict(X_test)))

#----output----
Train - Accuracy : 0.9809523809523809
Train - Confusion matrix : [[34  0  0]
                            [ 0 30  2]
                            [ 0  0 39]]
Train - classification report :
              precision   recall  f1-score   support
           0       1.00     1.00      1.00        34
           1       1.00     0.94      0.97        32
           2       0.95     1.00      0.97        39

   micro avg       0.98     0.98      0.98       105
   macro avg       0.98     0.98      0.98       105
weighted avg       0.98     0.98      0.98       105
```

```
Test - Accuracy : 0.9777777777777777
Test - Confusion matrix : [[16  0  0]
                           [ 0 17  1]
                           [ 0  0 11]]
Test - classification report :
              precision   recall  f1-score   support

           0       1.00     1.00      1.00        16
           1       1.00     0.94      0.97        18
           2       0.92     1.00      0.96        11

   micro avg       0.98     0.98      0.98        45
   macro avg       0.97     0.98      0.98        45
weighted avg       0.98     0.98      0.98        45
```

Generalized Linear Models

Generalized linear model (GLM) was an effort by John Nelder and Robert Wedderburn to unify commonly used various statistical models such as linear, logistic, Poisson, etc. (Table 3-12). Refer to Listing 3-37 for example code implementation.

Table 3-12. *Different GLM Distribution Family*

Family	Description
Binomial	Target variable is binary response
Poisson	Target variable is a count of occurrence
Gaussian	Target variable is a continuous number
Gamma	This distribution arises when the waiting times between Poisson distribution events are relevant (i.e., a number of events occurred between two time periods).
Inverse Gaussian	The tails of the distribution decrease slower than normal distribution (i.e., there is an inverse relationship between the time required to cover a unit distance and distance covered in unit time).
Negative Binomial	Target variable denotes the number of successes in a sequence before a random failure

Listing 3-37. Generalized Linear Model

```
df = pd.read_csv('Data/Grade_Set_1.csv')

print('####### Linear Regression Model ########')
# Create linear regression object
lr = lm.LinearRegression()

x= df.Hours_Studied[:, np.newaxis] # independent variable
y= df.Test_Grade.values            # dependent variable

# Train the model using the training sets
lr.fit(x, y)

print ("Intercept: ", lr.intercept_)
print ("Coefficient: ", lr.coef_)

print('\n####### Generalized Linear Model ########')
import statsmodels.api as sm

# To be able to run GLM, we'll have to add the intercept constant to x
variable
x = sm.add_constant(x, prepend=False)

# Instantiate a gaussian family model with the default link function.
model = sm.GLM(y, x, family = sm.families.Gaussian())
model = model.fit()
print (model.summary())
#----output----

####### Linear Regression Model ########
Intercept:  49.6777777778
Coefficient:  [ 5.01666667]
```

```
####### Generalized Linear Model ########
                    Generalized Linear Model Regression Results
==============================================================================
Dep. Variable:                     y   No. Observations:                    9
Model:                           GLM   Df Residuals:                        7
Model Family:               Gaussian   Df Model:                            1
Link Function:              identity   Scale:                          5.3627
Method:                         IRLS   Log-Likelihood:                -19.197
Date:                Sat, 09 Feb 2019  Deviance:                       37.539
Time:                       10:01:22   Pearson chi2:                     37.5
No. Iterations:                    3   Covariance Type:             nonrobust
==============================================================================
                 coef    std err          z      P>|z|      [0.025      0.975]
------------------------------------------------------------------------------
x1             5.0167      0.299     16.780      0.000       4.431       5.603
const         49.6778      1.953     25.439      0.000      45.850      53.505
==============================================================================
```

Note that the coefficients are the same for both linear regression and GLM. However, GLM can be used for other distributions such as binomial, Poisson, etc. by just changing the family parameter.

Supervised Learning–Process Flow

At this point, you have seen how to build a regression and a logistic regression model, so let me summarize the process flow for supervised learning in Figure 3-13.

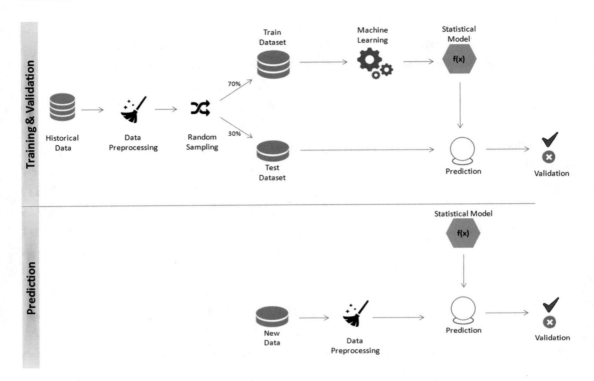

Figure 3-13. *Supervised learning process flow*

First, you need to train and validate a supervised model by applying ML techniques to historical data. Then apply this model to the new data set to predict the future value.

Decision Trees

In 1986, J.R. Quinlan published "Induction of Decision Trees" summarizing an approach to synthesizing decision trees using ML. It used an illustrative example dataset, where the objective was to make a decision on whether to play outside on a Saturday morning. As the name suggests, a decision tree is a tree-like structure where an internal node represents test on an attribute, each branch represents the outcome of the test, each leaf node represents the class label, and a decision is made after computing all attributes. A path from the root to a leaf represents classification rules. Thus, a decision tree consists of three types of nodes:

- Root node

- Branch node

- Leaf node (class label)

A decision tree model output is easy to interpret, and it provides the rules that drive a decision or event. In the preceding use case we can get the rules that lead to a don't play scenario: sunny and temperature >30°30⁰c, and rainy and windy are true. Often, business might be more interested in these decision rules than the decision itself. For example, an insurance company might be more interested in the rules or condition in which an insurance applicant should be sent for a medical checkup rather than feeding the applicants data to a black box model to find the decision (Figure 3-14).

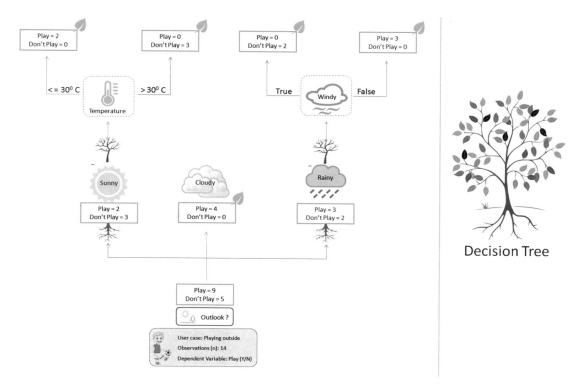

Figure 3-14. *J.R. Quinlan's example for synthesizing a decision tree*

Use training data to build the tree generator model, which will determine which variable to split at a node and the value of the split. The decision to stop or split again assigns leaf nodes to a class. The advantage of the decision tree is that there is no need for exclusive creation of dummy variables.

How the Tree Splits and Grows

- The base algorithm is known as a greedy algorithm, in which a tree is constructed in a top-down recursive divide-and-conquer manner.

- At the start, all the training examples are at the root.

- Input data is partitioned recursively based on selected attributes.

- Test attributes at each node are selected on the basis of a heuristic or statistical impurity measure example: Gini or information gain (entropy).

  - Gini = $1 - \sum_i (p_i)^2$, where p_i is the probability of each label.

  - Entropy = $-p \log2 (p) - q \log2 (q)$, where p and q represent the probability of success/failure respectively in a given node.

Conditions for Stopping Partitioning

- All samples for a given node belong to the same class.

- There are no remaining attributes for further partitioning—majority voting is employed for classifying the leaf

- There are no samples left.

Note   Default criterion is "Gini" because it's comparatively faster to compute than "entropy"; however, both measures give an almost identical decision on the split.

Listing 3-38 provides an example decision tree model implementation on the Iris data set.

Listing 3-38. Decision Tree Model

```
from sklearn import datasets
import numpy as np
import pandas as pd
from sklearn import tree
```

```
iris = datasets.load_iris()

# X = iris.data[:, [2, 3]]
X = iris.data
y = iris.target
from sklearn.preprocessing import StandardScaler

sc = StandardScaler()
sc.fit(X)
X = sc.transform(X)

# split data into train and test
from sklearn.model_selection import train_test_split

X_train, X_test, y_train, y_test = train_test_split(X, y, test_size=0.3,
random_state=0)

clf = tree.DecisionTreeClassifier(criterion = 'entropy', random_state=0)
clf.fit(X_train, y_train)

# generate evaluation metrics
print("Train - Accuracy :", metrics.accuracy_score(y_train, clf.predict
(X_train)))
print("Train - Confusion matrix :",metrics.confusion_matrix(y_train, clf.
predict(X_train)))
print("Train - classification report :", metrics.classification_report
(y_train, clf.predict(X_train)))

print("Test - Accuracy :", metrics.accuracy_score(y_test, clf.predict
(X_test)))
print("Test - Confusion matrix :",metrics.confusion_matrix(y_test, clf.
predict(X_test)))
print("Test - classification report :", metrics.classification_report
(y_test, clf.predict(X_test)))

tree.export_graphviz(clf, out_file='tree.dot')

from sklearn.externals.six import StringIO
import pydot
out_data = StringIO()
```

```
tree.export_graphviz(clf, out_file=out_data,
                     feature_names=iris.feature_names,
                     class_names=clf.classes_.astype(int).astype(str),
                     filled=True, rounded=True,
                     special_characters=True,
                     node_ids=1,)
graph = pydot.graph_from_dot_data(out_data.getvalue())
graph[0].write_pdf("iris.pdf")  # save to pdf
#----output----
Train - Accuracy : 1.0
Train - Confusion matrix : [[34  0  0]
                            [ 0 32  0]
                            [ 0  0 39]]
Train - classification report :
```

	precision	recall	f1-score	support
0	1.00	1.00	1.00	34
1	1.00	1.00	1.00	32
2	1.00	1.00	1.00	39
micro avg	1.00	1.00	1.00	105
macro avg	1.00	1.00	1.00	105
weighted avg	1.00	1.00	1.00	105

```
Test - Accuracy : 0.9777777777777777
Test - Confusion matrix : [[16  0  0]
                           [ 0 17  1]
                           [ 0  0 11]]
Test - classification report :
```

	precision	recall	f1-score	support
0	1.00	1.00	1.00	16
1	1.00	0.94	0.97	18
2	0.92	1.00	0.96	11
micro avg	0.98	0.98	0.98	45
macro avg	0.97	0.98	0.98	45
weighted avg	0.98	0.98	0.98	45

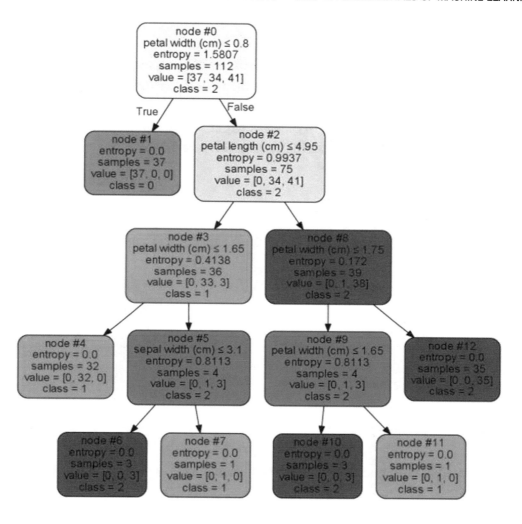

Key Parameters for Stopping Tree Growth

One of the key issues with a decision tree is that the tree can grow very large, ending up creating one leaf per observation.

max_features: Maximum features to be considered while deciding each split; default = "None," which means all features will be considered.

min_samples_split: Split will not be allowed for nodes that do not meet this number.

min_samples_leaf: Leaf node will not be allowed for nodes less than the minimum samples.

max_depth: No further split will be allowed; default = "None."

Support Vector Machine

Vladimir N. Vapnik and Alexey Ya. Chervonenkis in 1963 proposed SVM. A key objective of SVM is to draw a hyperplane that separates the two classes optimally such that the margin is maximum between the hyperplane and the observations. Figure 3-15 illustrates that there is a possibility of different hyperplanes. However, the objective of SVM is to find the one that gives us a high margin (Figure 3-15).

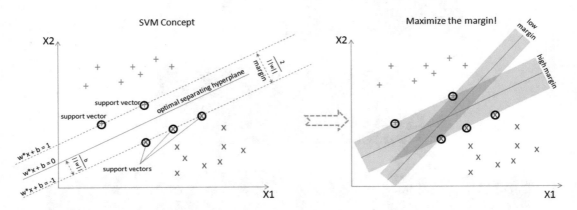

Figure 3-15. *Support vector machine*

To maximize the margin, we need to minimize $(1/2)\|w\|2$ subject to $yi(WTXi + b)-1 \geq 0$ for all i.

The final SVM equation can be written mathematically as

$$L = \sum_{i} d_i - \frac{1}{2}\sum_{ij}\alpha_i\alpha_j y_i y_j (\bar{X} i \bar{X} j)$$

Note   SVM is comparatively less prone to outliers than logistic regression, as it only cares about the points that are closest to the decision boundary or support vectors.

Key Parameters

C: This is the penalty parameter and helps in fitting the boundaries smoothly and appropriately; default = 1.

Kernel: A kernel is a similarity function for pattern analysis. It must be one of rbf/linear/poly/sigmoid/precomputed; default = "rbf" (radial basis function). Choosing the appropriate kernel will result in a better model fit (Listing 3-39).

Listing 3-39. Support Vector Machine (SVM) Model

```
from sklearn import datasets
import numpy as np
import pandas as pd
from sklearn import tree
from sklearn import metrics

iris = datasets.load_iris()

X = iris.data[:, [2, 3]]
y = iris.target

print('Class labels:', np.unique(y))
from sklearn.preprocessing import StandardScaler

sc = StandardScaler()
sc.fit(X)
X = sc.transform(X)

# split data into train and test
from sklearn.model_selection import train_test_split

X_train, X_test, y_train, y_test = train_test_split(X, y, test_size=0.3,
random_state=0)
from sklearn.svm import SVC

clf = SVC(kernel='linear', C=1.0, random_state=0)
clf.fit(X_train, y_train)

# generate evaluation metrics
# generate evaluation metrics
print("Train - Accuracy :", metrics.accuracy_score(y_train, clf.predict
(X_train)))
print("Train - Confusion matrix :",metrics.confusion_matrix(y_train, clf.
predict(X_train)))
print("Train - classification report :", metrics.classification_report
(y_train, clf.predict(X_train)))
```

```
print("Test - Accuracy :", metrics.accuracy_score(y_test, clf.predict
(X_test)))
print("Test - Confusion matrix :", metrics.confusion_matrix(y_test, clf.
predict(X_test)))
print("Test - classification report :", metrics.classification_report
(y_test, clf.predict(X_test)))
#----output----
Train - Accuracy : 0.9523809523809523
Train - Confusion matrix : [[34  0  0]
                            [ 0 30  2]
                            [ 0  3 36]]
Train - classification report :
```

	precision	recall	f1-score	support
0	1.00	1.00	1.00	34
1	0.91	0.94	0.92	32
2	0.95	0.92	0.94	39
micro avg	0.95	0.95	0.95	105
macro avg	0.95	0.95	0.95	105
weighted avg	0.95	0.95	0.95	105

```
Test - Accuracy : 0.9777777777777777
Test - Confusion matrix : [[16  0  0]
                           [ 0 17  1]
                           [ 0  0 11]]
Test - classification report :
```

	precision	recall	f1-score	support
0	1.00	1.00	1.00	16
1	1.00	0.94	0.97	18
2	0.92	1.00	0.96	11
micro avg	0.98	0.98	0.98	45
macro avg	0.97	0.98	0.98	45
weighted avg	0.98	0.98	0.98	45

Plotting decision boundary: Let's consider a two-class example to keep things simple (Listing 3-40).

Listing 3-40. Ploting SVM Decision Boundaries

```
# Let's use sklearn make_classification function to create some test data.
from sklearn.datasets import make_classification
X, y = make_classification(100, 2, 2, 0, weights=[.5, .5], random_state=0)

# build a simple logistic regression model
clf = SVC(kernel='linear', random_state=0)
clf.fit(X, y)

# get the separating hyperplane
w = clf.coef_[0]
a = -w[0] / w[1]
xx = np.linspace(-5, 5)
yy = a * xx - (clf.intercept_[0]) / w[1]

# plot the parallels to the separating hyperplane that pass through the
# support vectors
b = clf.support_vectors_[0]
yy_down = a * xx + (b[1] - a * b[0])
b = clf.support_vectors_[-1]
yy_up = a * xx + (b[1] - a * b[0])

# Plot the decision boundary
plot_decision_regions(X, y, classifier=clf)

# plot the line, the points, and the nearest vectors to the plane
plt.scatter(clf.support_vectors_[:, 0], clf.support_vectors_[:, 1], s=80,
facecolors='none')
plt.plot(xx, yy_down, 'k--')
plt.plot(xx, yy_up, 'k--')

plt.xlabel('X1')
plt.ylabel('X2')
plt.legend(loc='upper left')
plt.tight_layout()
plt.show()
#----output----
```

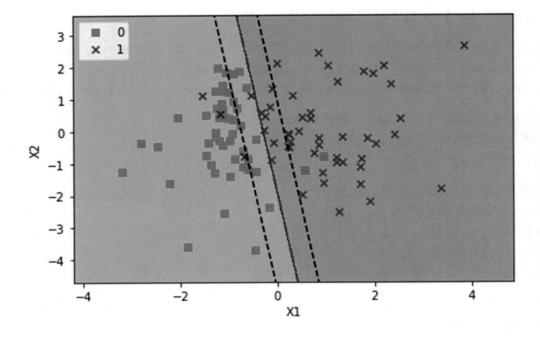

k-Nearest Neighbors

The k-nearest neighbor classification (kNN) was developed from the need to perform discriminant analysis when reliable parametric estimates of probability densities are unknown or difficult to determine. Fix and Hodges in 1951 introduced a nonparametric method for pattern classification that has since become known as the k-nearest neighbor rule.

As the name suggests, the algorithm works based on majority vote of its k-nearest neighbors class. In Figure 3-16, k = 5 nearest neighbors for the unknown data point are identified based on the chosen distance measure, and the unknown point will be classified based on majority class among identified nearest data points class. The key drawback of kNN is the complexity in searching the nearest neighbors for each sample (Listing 3-41).

Things to remember:

- Choose an odd k value for a two-class problem.

- k must not be a multiple of the number of classes.

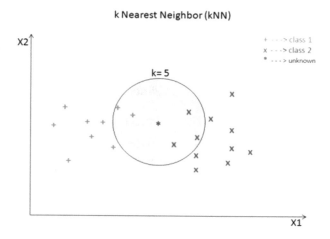

Figure 3-16. *k nearest neighbor with k = 5*

Listing 3-41. k Nearest Neighbor Model

```
from sklearn.neighbors import KNeighborsClassifier

clf = KNeighborsClassifier(n_neighbors=5, p=2, metric='minkowski')
clf.fit(X_train, y_train)

# generate evaluation metrics
print("Train - Accuracy :", metrics.accuracy_score(y_train, clf.predict
(X_train)))
print("Train - Confusion matrix :",metrics.confusion_matrix(y_train, clf.
predict(X_train)))
print("Train - classification report :", metrics.classification_report
(y_train, clf.predict(X_train)))

print("Test - Accuracy :", metrics.accuracy_score(y_test, clf.predict
(X_test)))
print("Test - Confusion matrix :", metrics.confusion_matrix(y_test, clf.
predict(X_test)))
```

```
print("Test - classification report :", metrics.classification_report
(y_test, clf.predict(X_test)))
#----output----
Train - Accuracy : 0.9714285714285714
Train - Confusion matrix : [[34  0  0]
                            [ 0 31  1]
                            [ 0  2 37]]
Train - classification report :
```

	precision	recall	f1-score	support
0	1.00	1.00	1.00	34
1	0.94	0.97	0.95	32
2	0.97	0.95	0.96	39
micro avg	0.97	0.97	0.97	105
macro avg	0.97	0.97	0.97	105
weighted avg	0.97	0.97	0.97	105

```
Test - Accuracy : 0.9777777777777777
Test - Confusion matrix : [[16  0  0]
                           [ 0 17  1]
                           [ 0  0 11]]
Test - classification report :
```

	precision	recall	f1-score	support
0	1.00	1.00	1.00	16
1	1.00	0.94	0.97	18
2	0.92	1.00	0.96	11
micro avg	0.98	0.98	0.98	45
macro avg	0.97	0.98	0.98	45
weighted avg	0.98	0.98	0.98	45

Note   Decision trees, SVM, and kNN-based algorithm concepts can essentially be applied to predict dependent variables, which are continuous numbers in nature; and Scikit-learn provides DecisionTreeRegressor, SVR (support vector regressor), and kNeighborsRegressor for the same.

Time-Series Forecasting

In simple terms, a series of data points that are collected sequentially at a regular interval over a time period is termed "time-series data." Time-series data having the mean and variance as constant is called a stationary time series.

Time series tend to have a linear relationship between lagged variables and this is called autocorrelation. Hence time-series historic data can be modeled to forecast future data points without the involvement of any other independent variables; these types of models are generally known as time-series forecasting. Some key areas of applications of time series are: sales forecasting, economic forecasting, and stock market forecasting.

Components of Time Series

A time series can be made up of three key components (Figure 3-17).

- *Trend*: A long-term increase or decrease is termed a trend.

- *Seasonality*: An effect of seasonal factors for a fixed or known period. For example, retail stores' sales will be high during weekends and holiday seasons.

- *Cycle*: These are the longer ups and downs that are not of fixed or known period, caused by external factors.

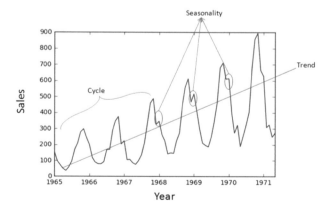

Figure 3-17. *Time series components*

Listing 3-42 shows the code implementation for decomposing the time-series components.

Autoregressive Integrated Moving Average (ARIMA)

ARIMA is a key and popular time-series model, so understanding the concept involved will set the basis for you around time-series modeling.

Autoregressive model (AM): As the name indicates, it is a regression of the variable against itself (i.e., the linear combination of past values of the variable is used to forecast the future value).

$y_t = c + \Phi_1 y_{t-1} + \Phi_2 y_{t-2} + \ldots + \Phi_n y_{t-n} + e_t$, where c is constant, e_t is the random error, y_{t-1} is first-order correlation, and y_{t-2} is second-order correlation between values two periods apart.

Moving average (MA): Instead of past values, past forecast errors are used to build a model.

$y_t = c + \theta y_{t-1} + \theta_2 y_{t-2} + \ldots + \theta_n y_{t-n} + e_t$

An autoregressive (AR), moving average (MA) model with integration (opposite of differencing) is called the ARIMA model.

$y_t = c + \phi_1 y_{t-1} + \phi_2 y_{t-2} + \ldots + \phi_n y_{t-n} + \theta y_{t-1} + \theta_2 y_{t-2} + \ldots + \theta_n y_{t-n} + e_t$

The predictors on the right side of the equation (p, d, q) are the lagged values, errors. These are the key parameters of ARIMA, and picking the right value for p, d, and q will yield better model results.

p = order of the autoregressive part. That is the number of unknown terms that multiply your signal at past times (so many past times as your value p).

d = degree of first differencing involved. The number of times you have to differentiate your time series to have a stationary one.

q = order of the moving average part. That is the number of unknown terms that multiply your forecast errors at past times (so many past times as your value q).

Running ARIMA Model

- Plot the chart to ensure trend, cycle, or seasonality exists in the data set.

- *Stationarize series*: To stationarize a series, we need to remove trend (varying mean) and seasonality (variance) components from the series. The moving average and differencing technique can be used to stabilize a trend, whereas log transform will stabilize the seasonality variance. Further, the Dickey Fuller test can be used to

assess the stationarity of a series, that is, the null hypothesis for a
Dickey Fuller test is that the data are stationary, so a test result with
p-value >0.05 means data is nonstationary (Listing 3-43).

- *Find optimal parameter*: Once the series is stationarized you can look
  at the autocorrelation function (ACF) and partial autocorrelation
  function (PACF) graphical plot to pick the number of AR or MA
  terms needed to remove autocorrelation. ACF is the bar chart
  between correlation coefficients and lags; similarly, PACF is the bar
  chart between partial correlation (correlation between variable and
  lag of itself did not explain bye correlation at all lower-order lags)
  coefficient and lags (Listing 3-44).

- *Build model and evaluate*: Since a time series is a continuous
  number, MAE and RMSE can be used to evaluate the deviation
  between actual and predicted values in the train data set. Other
  useful metrics would be the Akaike information criterion (AIC) and
  Bayesian information criterion (BIC). These are part of information
  theory to estimate the quality of individual models given a collection
  of models, and they favor a model with smaller residual errors
  (Listing 3-45).

AIC = -2log(L) + 2(p+q+k+1), where L is the maximum likelihood
function of a fitted model and p, q, k are the number of parameters in
the model

BIC = AIC+(log(T)−2)(p+q+k+1)

Listing 3-42. Decompose Time Series

```
# Data Source: O.D. Anderson (1976), in file: data/anderson14, Description:
Monthly sales of company X Jan '65 – May '71 C. Cahtfield
df = pd.read_csv('Data/TS.csv')
ts = pd.Series(list(df['Sales']), index=pd.to_
datetime(df['Month'],format='%Y-%m'))

from statsmodels.tsa.seasonal import seasonal_decompose
decomposition = seasonal_decompose(ts)
```

```
trend = decomposition.trend
seasonal = decomposition.seasonal
residual = decomposition.resid

plt.subplot(411)
plt.plot(ts_log, label='Original')
plt.legend(loc='best')
plt.subplot(412)
plt.plot(trend, label='Trend')
plt.legend(loc='best')
plt.subplot(413)
plt.plot(seasonal,label='Seasonality')
plt.legend(loc='best')
plt.tight_layout()
```

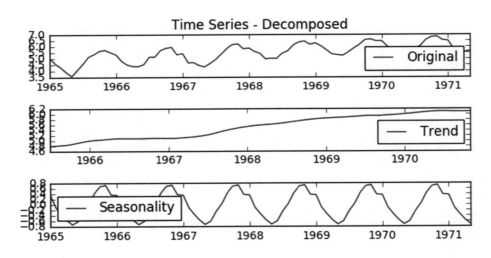

Checking for Stationary

Listing 3-43. Check Stationary

```
from statsmodels.tsa.stattools import adfuller

# log transform
ts_log = np.log(ts)
```

```python
ts_log.dropna(inplace=True)

s_test = adfuller(ts_log, autolag='AIC')
print ("Log transform stationary check p value: ", s_test[1])

#Take first difference:
ts_log_diff = ts_log - ts_log.shift()
ts_log_diff.dropna(inplace=True)
plt.title('Trend removed plot with first order difference')
plt.plot(ts_log_diff)
plt.ylabel('First order log diff')

s_test = adfuller(ts_log_diff, autolag='AIC')
print ("First order difference stationary check p value: ", s_test[1] )

# moving average smoothens the line
moving_avg = ts_log.rolling(12).mean()

fig, (ax1, ax2) = plt.subplots(1, 2, figsize = (10,3))
ax1.set_title('First order difference')
ax1.tick_params(axis='x', labelsize=7)
ax1.tick_params(axis='y', labelsize=7)
ax1.plot(ts_log_diff)

ax2.plot(ts_log)
ax2.set_title('Log vs Moving AVg')
ax2.tick_params(axis='x', labelsize=7)
ax2.tick_params(axis='y', labelsize=7)
ax2.plot(moving_avg, color='red')
plt.tight_layout()

#----output----
Log transform stationary check p value:  0.785310212485
First order difference stationary check p value:  0.0240253928399
```

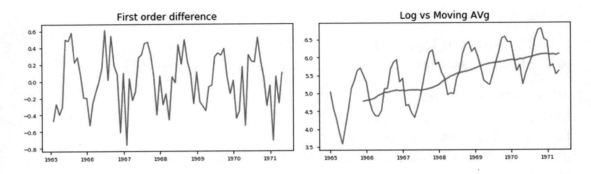

Autocorrelation Test

We determined that the log of time series requires at least one order different to stationarize. Now let's plot ACF and PACF charts for first order log series.

Listing 3-44. Check Autocorrelation

```
fig, (ax1, ax2) = plt.subplots(1, 2, figsize = (10,3))

# ACF chart
fig = sm.graphics.tsa.plot_acf(ts_log_diff.values.squeeze(), lags=20,
ax=ax1)

# draw 95% confidence interval line
ax1.axhline(y=-1.96/np.sqrt(len(ts_log_diff)),linestyle='--',color='gray')
ax1.axhline(y=1.96/np.sqrt(len(ts_log_diff)),linestyle='--',color='gray')
ax1.set_xlabel('Lags')

# PACF chart
fig = sm.graphics.tsa.plot_pacf(ts_log_diff, lags=20, ax=ax2)

# draw 95% confidence interval line
ax2.axhline(y=-1.96/np.sqrt(len(ts_log_diff)),linestyle='--',color='gray')
ax2.axhline(y=1.96/np.sqrt(len(ts_log_diff)),linestyle='--',color='gray')
ax2.set_xlabel('Lags')
#----output----
```

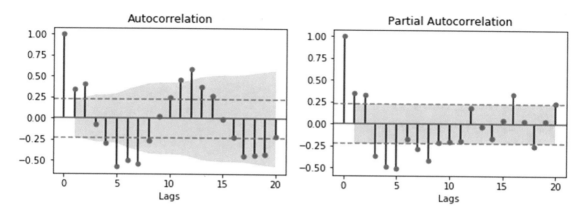

The PACF plot has a significant spike only at lag 1, meaning that all the higher order autocorrelations are effectively explained by the lag-1 and lag-2 autocorrelation. Ideal lag values are p = 2 and q = 2 (i.e., the lag value where the ACF/PACF chart crosses the upper confidence interval for the first time).

Build Model and Evaluate

Let's fit the ARIMA model on the dataset and evaluate the model performance.

Listing 3-45. Build ARIMA Model and Evaluate

```
# build model
model = sm.tsa.ARIMA(ts_log, order=(2,0,2))
results_ARIMA = model.fit(disp=-1)

ts_predict = results_ARIMA.predict()

# Evaluate model
print("AIC: ", results_ARIMA.aic)
print("BIC: ", results_ARIMA.bic)

print("Mean Absolute Error: ", mean_absolute_error(ts_log.values,
ts_predict.values))
print("Root Mean Squared Error: ", np.sqrt(mean_squared_error(ts_log.
values, ts_predict.values)))
```

```
# check autocorrelation
print("Durbin-Watson statistic :", sm.stats.durbin_watson(results_ARIMA.
resid.values))
#----output-----
AIC:  7.8521105380873735
BIC:  21.914943069209478
Mean Absolute Error:  0.19596606887750853
Root Mean Squared Error:  0.2397921908617542
Durbin-Watson statistic : 1.8645776109746208
```

The usual practice is to build several models with different p and q and select the one with the smallest value of AIC, BIC, MAE, and RMSE. Now lets' increase p to 3 and see if there is any difference in the result (Listing 3-46).

Listing 3-46. Build ARIMA Model and Evaluate by Increasing p to 3

```
model = sm.tsa.ARIMA(ts_log, order=(3,0,2))
results_ARIMA = model.fit(disp=-1)

ts_predict = results_ARIMA.predict()
plt.title('ARIMA Prediction - order(3,0,2)')
plt.plot(ts_log, label='Actual')
plt.plot(ts_predict, 'r--', label='Predicted')
plt.xlabel('Year-Month')
plt.ylabel('Sales')
plt.legend(loc='best')

print("AIC: ", results_ARIMA.aic)
print("BIC: ", results_ARIMA.bic)

print("Mean Absolute Error: ", mean_absolute_error(ts_log.values,
ts_predict.values))
print("Root Mean Squared Error: ", np.sqrt(mean_squared_error(ts_log.
values, ts_predict.values)))

# check autocorrelation
print("Durbin-Watson statistic :", sm.stats.durbin_watson(results_ARIMA.
resid.values))
AIC:   -7.786042455163056
```

BIC: 8.620595497812733
Mean Absolute Error: 0.16721947678957297
Root Mean Squared Error: 0.21618486190507652
Durbin-Watson statistic : 2.5184568082461936

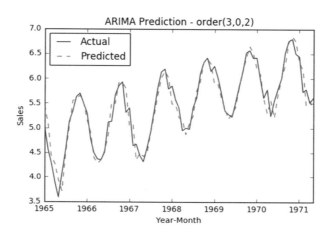

Let's try it with first order differencing (i.e., d = 1) to see if the model performance improves (Listing 3-47).

Listing 3-47. ARIMA with First Order Differencing

```
model = sm.tsa.ARIMA(ts_log, order=(3,1,2))
results_ARIMA = model.fit(disp=-1)

ts_predict = results_ARIMA.predict()

# Correctcion for difference
predictions_ARIMA_diff = pd.Series(ts_predict, copy=True)
predictions_ARIMA_diff_cumsum = predictions_ARIMA_diff.cumsum()
predictions_ARIMA_log = pd.Series(ts_log.ix[0], index=ts_log.index)
predictions_ARIMA_log = predictions_ARIMA_log.add(predictions_ARIMA_diff_
cumsum,fill_value=0)

#----output----
plt.title('ARIMA Prediction - order(3,1,2)')
plt.plot(ts_log, label='Actual')
plt.plot(predictions_ARIMA_log, 'r--', label='Predicted')
```

```
plt.xlabel('Year-Month')
plt.ylabel('Sales')
plt.legend(loc='best')

print("AIC: ", results_ARIMA.aic)
print("BIC: ", results_ARIMA.bic)

print("Mean Absolute Error: ", mean_absolute_error(ts_log_diff.values,
ts_predict.values))
print("Root Mean Squared Error: ", np.sqrt(mean_squared_error(ts_log_diff.
values, ts_predict.values)))

# check autocorrelation
print("Durbin-Watson statistic :", sm.stats.durbin_watson(results_ARIMA.
resid.values))

#----output----
AIC:   -35.41898773672588
BIC:   -19.103854354721562
Mean Absolute Error:   0.13876538862134086
Root Mean Squared Error:   0.1831024379477494
Durbin-Watson statistic : 1.941165833847913
```

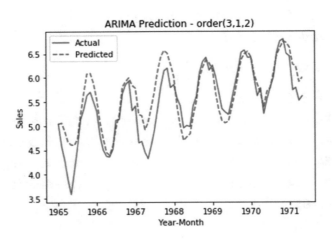

In the preceding chart, we can see that the model is overpredicting at some places, and AIC and BIC values are higher than the previous model. Note: AIC/BIC can be positive or negative; however, we should look at the absolute value of it for evaluation.

Predicting Future Values

Below values (p=3, d=0, q=2) is giving the smaller number for evaluation metrics, so let's use this as a final model to predict the future values, for the year 1972 (Listing 3-48).

Listing 3-48. ARIMA Predict Function

```
# final model
model = sm.tsa.ARIMA(ts_log, order=(3,0,2))
results_ARIMA = model.fit(disp=-1)

# predict future values
ts_predict = results_ARIMA.predict('1971-06-01', '1972-05-01')
plt.title('ARIMA Future Value Prediction - order(3,1,2)')
plt.plot(ts_log, label='Actual')
plt.plot(ts_predict, 'r--', label='Predicted')
plt.xlabel('Year-Month')
plt.ylabel('Sales')
plt.legend(loc='best')
#----output----
```

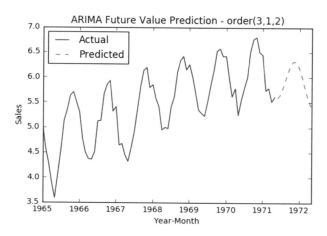

Note   A minimum of 3 to 4 years' worth of historical data is required to ensure the seasonal patterns are regular.

Unsupervised Learning Process Flow

The unsupervised learning process flow is shown in Figure 3-18 below. Similar to supervised learning, we can train a model and use it to predict the unknown data set. However, the key difference is that there is no predefined category or labels available for a target variable, and the goal often is to create a category or label based on patterns available in data.

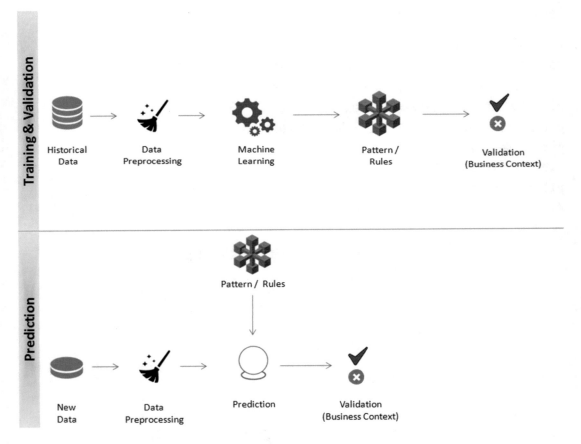

Figure 3-18. *Unsupervised learning process flow*

Clustering

Clustering is an unsupervised learning problem. The key objective is to identify distinct groups (called clusters) based on some notion of similarity within a given dataset. Clustering analysis origins can be traced to the areas of anthropology and psychology in the 1930s. Most popularly used clustering techniques are k-means (divisive) and hierarchical (agglomerative).

K-means

The key objective of the K-means algorithm is to organize data into clusters such that there is a high intracluster similarity and low intercluster similarity. An item will only belong to one cluster, not several (i.e., it generates a specific number of disjoint, non-hierarchical clusters). K-means uses the strategy of divide and concur, and is a classic example for expectation maximization (EM) algorithms. EM algorithms are made up of two steps: the first step, known as the expectation (E), is to find the expected point associated with a cluster; the second step, known as maximization (M), is to improve the estimation of the cluster using knowledge from the first step. The two steps are processed repeatedly until convergence is reached.

Suppose we have "n" data points that we need to cluster into k (c1, c2, c3) groups (Figure 3-19).

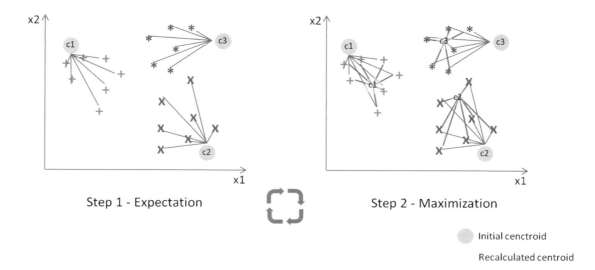

Figure 3-19. *Expectation maximization algorithm workflow*

Step 1: In the first step, k centroids (in the preceding case k = 3) is randomly picked (only in the first iteration) and all the points that are nearest to each centroid point are assigned to that specific cluster. The centroid is the arithmetic mean or average position of all the points.

Step 2: Here the centroid point is recalculated using the average of the coordinates of all the points in that cluster. Then step one is repeated (assign the nearest point) until the clusters converge.

Note: K-means is designed for Euclidean distance only.

$$Euclidean\,Distance = d = \sqrt{\sum_{i=1}^{N}(X_i - Y_i)^2}$$

Limitations of K-means

- K-means clustering needs the number of clusters to be specified.

- K-means has problems when clusters are of differing sizes, densities, and nonglobular shapes.

- The presence of an outlier can skew the results.

Let's load the Iris data and assume for a moment that the species column is missing—we have the measured values for sepal length/width and petal length/width but we do not know how many species exist.

Now let's use unsupervised learning (clustering) to find out how many species exist. The goal here is to group all similar items into a cluster. We can assume a k of 3 for now; we'll learn later about the approach to finding the value of k (Listing 3-49).

Listing 3-49. k-means Clustering

```
from sklearn import datasets
import numpy as np
import pandas as pd
from sklearn.cluster import KMeans
from sklearn.preprocessing import StandardScaler

iris = datasets.load_iris()

# Let's convert to dataframe
```

```
iris = pd.DataFrame(data= np.c_[iris['data'], iris['target']],
                    columns= iris['feature_names'] + ['species'])

# let's remove spaces from column name
iris.columns = iris.columns.str.replace(' ',")
iris.head()

X = iris.ix[:,:3]  # independent variables
y = iris.species    # dependent variable
sc = StandardScaler()
sc.fit(X)
X = sc.transform(X)

# K Means Cluster
model = KMeans(n_clusters=3, random_state=11)
model.fit(X)
print (model.labels_)
# ----output----
[1 1 1 1 1 1 1 1 1 1 1 1 1 1 1 1 1 1 1 1 1 1 1 1 1 1 1 1 1 1 1 1 1 1 1 1 1
 1 1 1 1 0 1 1 1 1 1 1 1 1 2 2 2 0 2 0 2 0 2 0 0 0 0 0 0 2 0 0 0 0 2 0 0 0
 2 2 2 2 0 0 0 0 0 0 0 2 2 0 0 0 0 2 0 0 0 0 0 0 0 0 2 0 2 2 2 2 0 2 2 2 2
 2 2 0 0 2 2 2 2 0 2 0 2 0 2 2 0 2 2 2 2 2 2 2 0 2 2 2 0 2 2 2 0 2 2 2 0 2
 2 0]]
```

We see that the clustering algorithm has assigned a cluster label for each record. Let's compare this with the actual species label to understand the accuracy of grouping similar records (Listing 3-50).

Listing 3-50. Accuracy of k-means Clustering

```
# since its unsupervised the labels have been assigned
# not in line with the actual lables so let's convert all the 1s to 0s and
0s to 1s
# 2's look fine
iris['pred_species'] = np.choose(model.labels_, [1, 0, 2]).astype
(np.int64)
```

```
print ("Accuracy :", metrics.accuracy_score(iris.species, iris.pred_
species))
print ("Classification report :", metrics.classification_report(iris.
species, iris.pred_species))

# Set the size of the plot
plt.figure(figsize=(10,7))

# Create a colormap for red, green and blue
cmap = ListedColormap(['r', 'g', 'b'])

# Plot Sepal
plt.subplot(2, 2, 1)
plt.scatter(iris['sepallength(cm)'], iris['sepalwidth(cm)'], c=cmap(iris.
species), marker='o', s=50)
plt.xlabel('sepallength(cm)')
plt.ylabel('sepalwidth(cm)')
plt.title('Sepal (Actual)')

plt.subplot(2, 2, 2)
plt.scatter(iris['sepallength(cm)'], iris['sepalwidth(cm)'], c=cmap(iris.
pred_species), marker='o', s=50)
plt.xlabel('sepallength(cm)')
plt.ylabel('sepalwidth(cm)')
plt.title('Sepal (Predicted)')

plt.subplot(2, 2, 3)
plt.scatter(iris['petallength(cm)'], iris['petalwidth(cm)'], c=cmap(iris.
species),marker='o', s=50)
plt.xlabel('petallength(cm)')
plt.ylabel('petalwidth(cm)')
plt.title('Petal (Actual)')

plt.subplot(2, 2, 4)
plt.scatter(iris['petallength(cm)'], iris['petalwidth(cm)'], c=cmap(iris.
pred_species),marker='o', s=50)
plt.xlabel('petallength(cm)')
plt.ylabel('petalwidth(cm)')
```

```
plt.title('Petal (Predicted)')
plt.tight_layout()
#----output----
Accuracy : 0.8066666666666666
Classification report :
```

	precision	recall	f1-score	support
0.0	1.00	0.98	0.99	50
1.0	0.71	0.70	0.71	50
2.0	0.71	0.74	0.73	50
micro avg	0.81	0.81	0.81	150
macro avg	0.81	0.81	0.81	150
weighted avg	0.81	0.81	0.81	150

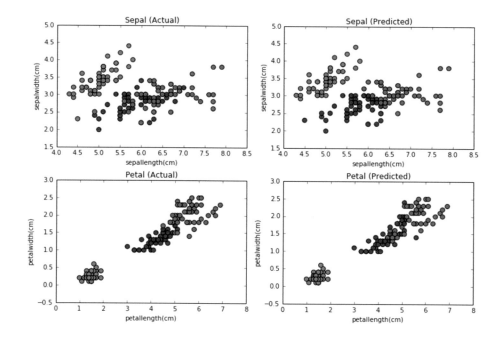

We can see from the preceding chart that K-means has done a decent job of clustering similar labels with an accuracy of 80% compared with the actual labels.

Finding the Value of k

Two methods are commonly used to determine the value of k:

- Elbow method

- Average silhouette method

Elbow Method

Perform K-means clustering on the dataset for a range of value k (for example 1 to 10) and calculate the SSE or percentage of variance explained for each k. Plot a line chart for cluster number vs. SSE, then look for an elbow shape on the line graph, which is the ideal number of clusters. With the increase in k, the SSE tends to decrease toward 0. The SSE is zero if is equal to the total number of data points in the dataset, as at this stage each data point becomes its own cluster, and no error exists between the cluster and its center. So the goal with the elbow method is to choose a small value of k that has a low SSE, and the elbow usually represents this value. Percentage of variance explained tends to increase with an increase in k, and we'll pick the point where the elbow shape appears (Listing 3-51).

Listing 3-51. Elbow Method

```
from scipy.spatial.distance import cdist, pdist
from sklearn.cluster import KMeans

K = range(1,10)
KM = [KMeans(n_clusters=k).fit(X) for k in K]
centroids = [k.cluster_centers_ for k in KM]

D_k = [cdist(X, cent, 'euclidean') for cent in centroids]
cIdx = [np.argmin(D,axis=1) for D in D_k]
dist = [np.min(D,axis=1) for D in D_k]
avgWithinSS = [sum(d)/X.shape[0] for d in dist]

# Total with-in sum of square
wcss = [sum(d**2) for d in dist]
tss = sum(pdist(X)**2)/X.shape[0]
bss = tss-wcss
varExplained = bss/tss*100
```

```
kIdx = 10-1
##### plot ###
kIdx = 2

# elbow curve
# Set the size of the plot
plt.figure(figsize=(10,4))

plt.subplot(1, 2, 1)
plt.plot(K, avgWithinSS, 'b*-')
plt.plot(K[kIdx], avgWithinSS[kIdx], marker='o', markersize=12,
    markeredgewidth=2, markeredgecolor='r', markerfacecolor='None')
plt.grid(True)
plt.xlabel('Number of clusters')
plt.ylabel('Average within-cluster sum of squares')
plt.title('Elbow for KMeans clustering')

plt.subplot(1, 2, 2)
plt.plot(K, varExplained, 'b*-')
plt.plot(K[kIdx], varExplained[kIdx], marker='o', markersize=12,
    markeredgewidth=2, markeredgecolor='r', markerfacecolor='None')
plt.grid(True)
plt.xlabel('Number of clusters')
plt.ylabel('Percentage of variance explained')
plt.title('Elbow for KMeans clustering')
plt.tight_layout()
#----output----
```

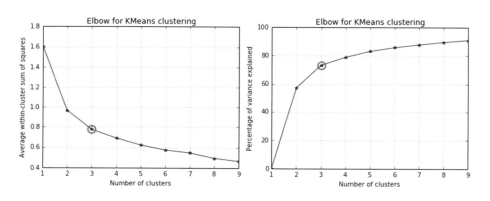

Average Silhouette Method

In 1986, Peter J. Rousseeuw described the silhouette method, which aims to explain the consistency within cluster data. Silhouette value will range between -1 and 1. A high value indicates that items are well matched within the cluster and weakly matched to the neighboring cluster (Listing 3-52).

s(i) = b(i) – a(i) / max {a(i), b(i)}, where a(i) is the average dissimilarity of the i-th item with other data points from the same cluster, and b(i) is the lowest average dissimilarity of i to other cluster to which i is not a member.

Listing 3-52. Silhouette Method

```
from sklearn.metrics import silhouette_samples, silhouette_score
from matplotlib import cm

score = []
for n_clusters in range(2,10):
    kmeans = KMeans(n_clusters=n_clusters)
    kmeans.fit(X)

    labels = kmeans.labels_
    centroids = kmeans.cluster_centers_

    score.append(silhouette_score(X, labels, metric='euclidean'))

    # Set the size of the plot
plt.figure(figsize=(10,4))

plt.subplot(1, 2, 1)
plt.plot(score)
plt.grid(True)
plt.ylabel("Silouette Score")
plt.xlabel("k")
plt.title("Silouette for K-means")

# Initialize the clusterer with n_clusters value and a random generator
model = KMeans(n_clusters=3, init='k-means++', n_init=10, random_state=0)
model.fit_predict(X)
```

```
cluster_labels = np.unique(model.labels_)
n_clusters = cluster_labels.shape[0]

# Compute the silhouette scores for each sample
silhouette_vals = silhouette_samples(X, model.labels_)

plt.subplot(1, 2, 2)

# Get spectral values for colormap.
cmap = cm.get_cmap("Spectral")

y_lower, y_upper = 0,0
yticks = []
for i, c in enumerate(cluster_labels):
    c_silhouette_vals = silhouette_vals[cluster_labels]
    c_silhouette_vals.sort()
    y_upper += len(c_silhouette_vals)
    color = cmap(float(i) / n_clusters)
    plt.barh(range(y_lower, y_upper), c_silhouette_vals, facecolor=color,
    edgecolor=color, alpha=0.7)
    yticks.append((y_lower + y_upper) / 2)
    y_lower += len(c_silhouette_vals)
silhouette_avg = np.mean(silhouette_vals)

plt.yticks(yticks, cluster_labels+1)

# The vertical line for average silhouette score of all the values
plt.axvline(x=silhouette_avg, color="red", linestyle="--")

plt.ylabel('Cluster')
plt.xlabel('Silhouette coefficient')
plt.title("Silouette for K-means")
plt.show()

#---output----
```

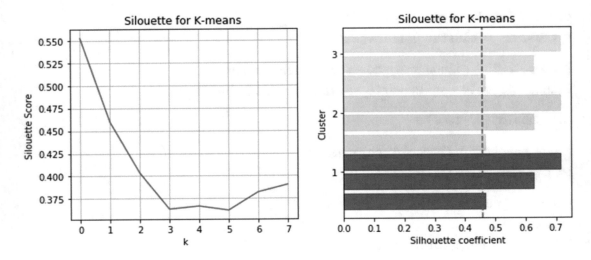

Hierarchical Clustering

Agglomerative clustering is a hierarchical cluster technique that builds nested clusters with a bottom-up approach where each data point starts in its own cluster and as we move up, the clusters are merged based on a distance matrix.

Key parameters

n_clusters: This is the number of clusters to find; the default is 2.

Linkage: It has to be one of the following: Ward's or complete or average; default = Ward (Figure 3-20).

Let's understand each linkage a bit more. The Ward's method will merge clusters if the in-cluster variance or the sum of square error is minimum. All pairwise distance of both clusters are used in the "average" method, and it is less affected by outliers. The "complete" method considers the distance between the farthest elements of two clusters, so it is also known as maximum linkage. Listing 3-53 is an example implementation code for hierarchical clustering.

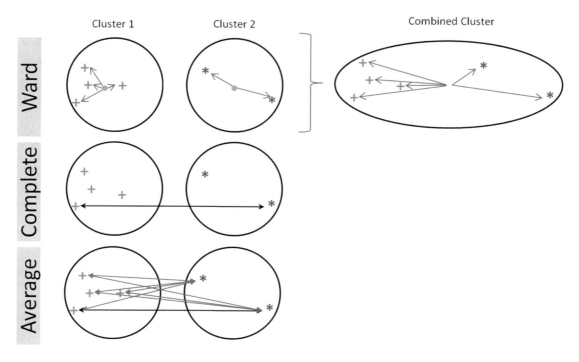

Figure 3-20. *Agglomerative clustering linkage*

Listing 3-53. Hierarchical Clustering

```
from sklearn.cluster import AgglomerativeClustering

# Agglomerative Cluster
model = AgglomerativeClustering(n_clusters=3)

# lets fit the model to the iris data set that we imported in Listing 3-49
model.fit(X)

print(model.labels_)
iris['pred_species'] =  model.labels_

print("Accuracy :", metrics.accuracy_score(iris.species, iris.pred_
species))
print("Classification report :", metrics.classification_report(iris.
species, iris.pred_species))
```

```
#----outout----
[0 0 0 0 0 0 0 0 0 0 0 0 0 0 0 0 0 0 0 0 0 0 0 0 0 0 0 0 0 0 0 0 0 0 0 0 0
 0 0 0 0 1 0 0 0 0 0 0 0 0 0 2 2 2 1 2 1 2 1 2 1 1 1 1 1 1 2 1 1 1 1 2 1 1 1
 1 2 2 2 1 1 1 1 1 1 1 2 2 1 1 1 1 1 1 1 1 1 1 1 1 2 1 2 1 2 2 1 2 1 2 2
 1 2 1 1 2 2 2 2 1 2 1 2 1 2 2 1 1 1 2 2 2 1 1 1 2 2 2 1 2 2 2 1 2 2 2 1 2
 2 1]
Accuracy : 0.7733333333333333
Classification report :
```

	precision	recall	f1-score	support
0.0	1.00	0.98	0.99	50
1.0	0.64	0.74	0.69	50
2.0	0.70	0.60	0.65	50
micro avg	0.77	0.77	0.77	150
macro avg	0.78	0.77	0.77	150
weighted avg	0.78	0.77	0.77	150

Hierarchical clusterings result arrangement can be better interpreted with dendogram visualization. SciPy provides necessary functions for dendogram visualization (Listing 3-54). Currently, Scikit-learn lacks these functions.

Listing 3-54. Hierarchical Clustering

```
from scipy.cluster.hierarchy import cophenet, dendrogram, linkage
from scipy.spatial.distance import pdist

# generate the linkage matrix
Z = linkage(X, 'ward')
c, coph_dists = cophenet(Z, pdist(X))

# calculate full dendrogram
plt.figure(figsize=(25, 10))
plt.title('Agglomerative Hierarchical Clustering Dendrogram')
plt.xlabel('sample index')
```

```
plt.ylabel('distance')
dendrogram(
    Z,
    leaf_rotation=90.,  # rotates the x axis labels
    leaf_font_size=8.,  # font size for the x axis labels
)
plt.tight_layout()
#----output----
```

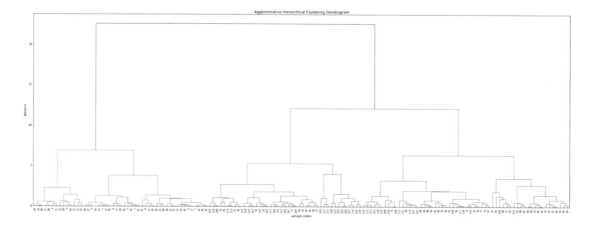

Since we know that k = 3, we can cut the tree at a distance threshold of around ten to get exactly three distinct clusters.

Principal Component Analysis (PCA)

The existence of a large number of features or dimension makes analysis computationally intensive and hard to perform ML tasks for pattern identification. Principal component analysis (PCA) is the most popular unsupervised linear transformation technique for dimensionality reduction. PCA finds the directions of maximum variance in high-dimensional data such that most of the information is retained, and projects it onto a smaller dimensional subspace (Figure 3-21).

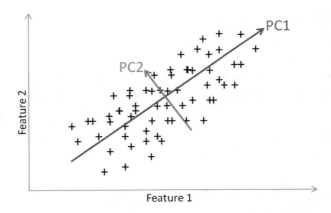

Figure 3-21. *Principal Component Analysis*

The PCA approach can be summarized as follows:

- Standardize data.

- Use standardized data to generate a covariance matrix or correlation matrix.

- Perform eigen decomposition: compute eigenvectors that are the principal component, which will give the direction, and compute eigenvalues, which will give the magnitude.

- Sort the eigen pairs and select eigenvectors with the largest eigenvalues, which cumulatively captures information above a certain threshold (say 95%).

Listing 3-55. Principal Component Analysis

```
from sklearn import datasets
from sklearn.preprocessing import StandardScaler
from sklearn.decomposition import PCA

iris = datasets.load_iris()
X = iris.data

# standardize data
X_std = StandardScaler().fit_transform(X)
```

```
# create covariance matrix
cov_mat = np.cov(X_std.T)

print('Covariance matrix \n%s' %cov_mat)

eig_vals, eig_vecs = np.linalg.eig(cov_mat)
print('Eigenvectors \n%s' %eig_vecs)
print('\nEigenvalues \n%s' %eig_vals)

# sort eigenvalues in decreasing order
eig_pairs = [(np.abs(eig_vals[i]), eig_vecs[:,i]) for i in range(len
(eig_vals))]

tot = sum(eig_vals)
var_exp = [(i / tot)*100 for i in sorted(eig_vals, reverse=True)]
print("Cummulative Variance Explained", cum_var_exp)

plt.figure(figsize=(6, 4))

plt.bar(range(4), var_exp, alpha=0.5, align='center',
        label='Individual explained variance')
plt.step(range(4), cum_var_exp, where='mid',
         label='Cumulative explained variance')
plt.ylabel('Explained variance ratio')
plt.xlabel('Principal components')
plt.legend(loc='best')
plt.tight_layout()
plt.show()
#----output----
Covariance matrix
[[ 1.00671141 -0.11835884  0.87760447  0.82343066]
 [-0.11835884  1.00671141 -0.43131554 -0.36858315]
 [ 0.87760447 -0.43131554  1.00671141  0.96932762]
 [ 0.82343066 -0.36858315  0.96932762  1.00671141]]
Eigenvectors
[[ 0.52106591 -0.37741762 -0.71956635  0.26128628]
 [-0.26934744 -0.92329566  0.24438178 -0.12350962]
 [ 0.5804131  -0.02449161  0.14212637 -0.80144925]
 [ 0.56485654 -0.06694199  0.63427274  0.52359713]]
```

```
Eigenvalues
[2.93808505 0.9201649  0.14774182 0.02085386]

Cummulative Variance Explained
[ 72.96244541  95.8132072    99.48212909 100.          ]
```

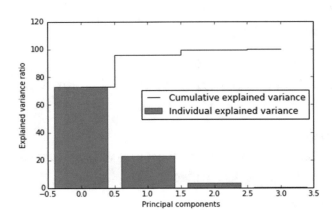

In the preceding plot, we can see that the first three principal components explain 99% of the variance. Let's perform PCA using Scikit-learn and plot the 1st three eigenvectors.

Listing 3-56 shows the code implementation example for visualizing the PCA.

Listing 3-56. Visualize PCA

```
# source: http://scikit-learn.org/stable/auto_examples/datasets/plot_iris_
dataset.html#
import matplotlib.pyplot as plt
from mpl_toolkits.mplot3d import Axes3D
from sklearn import datasets
from sklearn.decomposition import PCA

# import some data to play with
iris = datasets.load_iris()
Y = iris.target

# To getter a better understanding of interaction of the dimensions
# plot the first three PCA dimensions
fig = plt.figure(1, figsize=(8, 6))
```

```
ax = Axes3D(fig, elev=-150, azim=110)
X_reduced = PCA(n_components=3).fit_transform(iris.data)
ax.scatter(X_reduced[:, 0], X_reduced[:, 1], X_reduced[:, 2], c=Y,
cmap=plt.cm.Paired)
ax.set_title("First three PCA directions")
ax.set_xlabel("1st eigenvector")
ax.w_xaxis.set_ticklabels([])
ax.set_ylabel("2nd eigenvector")
ax.w_yaxis.set_ticklabels([])
ax.set_zlabel("3rd eigenvector")
ax.w_zaxis.set_ticklabels([])
plt.show()
#---output----
```

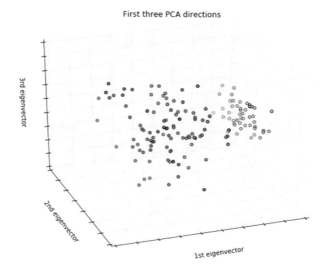

Summary

With this we have reached the end of step 2. We briefly learned different fundamental ML concepts and their implementation. Data quality is an important aspect to build efficient ML systems. In line with this we learned about different types of data, commonly practiced EDA techniques for understanding the data quality, and the fundamental preprocessing techniques to fix the data gaps. Supervised models such as linear and nonlinear regression techniques are useful to model patterns to predict continuous numerical data types. Whereas logistic regression, decision trees, SVM, and kNN are useful to model classification problems (functions are available to use for regression as well). We also learned about ARIMA, which is one of the key time-series forecasting models. Unsupervised techniques such as k-means and hierarchical clustering are useful to group similar items, whereas principal component analysis can be used to reduce large dimension data to lower dimension to enable efficient computation.

In the next step, you'll learn how to pick the best parameters for the model, widely known as "hyperparameter tuning" to improve model accuracy. What are the common practices followed to pick the best model among multiple models for a given problem? You'll also learn to combine multiple models to get the best from individual models.

CHAPTER 4

Step 4: Model Diagnosis and Tuning

In this chapter, we'll learn about the different pitfalls that one should be aware and will encounter while building a machine learning (ML) system. We'll also learn industry standard efficient design practices to solve it.

Throughout this chapter we'll mostly be using a dataset from the UCI repository, "Pima Indian diabetes," which has 768 records, 8 attributes, 2 classes, 268 (34.9%) positive results for a diabetes test, and 500 (65.1%) negative results. All patients were females at least 21 years old of Pima Indian heritage.

Attributes of the dataset:

1. Number of times pregnant

2. Plasma glucose concentration at 2 hours in an oral glucose tolerance test

3. Diastolic blood pressure (mmHg)

4. Triceps skin fold thickness (mm)

5. 2-Hour serum insulin (mu U/ml)

6. Body mass index (weight in kg/(height in m)^2)

7. Diabetes pedigree function

8. Age (years)

© Manohar Swamynathan 2019
M. Swamynathan, *Mastering Machine Learning with Python in Six Steps*,
https://doi.org/10.1007/978-1-4842-4947-5_4

Optimal Probability Cutoff Point

Predicted probability is a number between 0 and 1. Traditionally, >.5 is the cutoff point used for converting predicted probability to 1 (positive), otherwise 0 (negative). This logic works well when your training data set has an equal example of positive and negative cases; however, this is not the case in real-world scenarios.

The solution is to find the optimal cutoff point (i.e., the point where the true positive rate is high and the false positive rate is low). Anything above this threshold can be labeled as 1, else 0. Listing 4-1 should illustrate this point, so let's load the data and check the class distribution.

Listing 4-1. Load Data and Check the Class Distribution

```
import pandas as pd
import pylab as plt
import numpy as np

from sklearn.model_selection import train_test_split
from sklearn.linear_model import LogisticRegression
from sklearn import metrics

# read the data in
df = pd.read_csv("Data/Diabetes.csv")

# target variable % distribution
print (df['class'].value_counts(normalize=True))
#----output----
0    0.651042
1    0.348958
```

Let's build a quick logistic regression model and check the accuracy (Listing 4-2).

Listing 4-2. Build a Logistic Regression Model and Evaluate the Performance

```
X = df.ix[:,:8]       # independent variables
y = df['class']       # dependent variables

# evaluate the model by splitting into train and test sets
X_train, X_test, y_train, y_test = train_test_split(X, y, test_size=0.3,
random_state=0)
```

```
# instantiate a logistic regression model, and fit
model = LogisticRegression()
model = model.fit(X_train, y_train)

# predict class labels for the train set. The predict fuction converts
probability values > .5 to 1 else 0
y_pred = model.predict(X_train)

# generate class probabilities
# Notice that 2 elements will be returned in probs array,
# 1st element is probability for negative class,
# 2nd element gives probability for positive class
probs = model.predict_proba(X_train)
y_pred_prob = probs[:, 1]

# generate evaluation metrics
print ("Accuracy: ", metrics.accuracy_score(y_train, y_pred))
#----output----
Accuracy:   0.767225325885
```

The optimal cut-off would be where the true positive rate (tpr) is high and the false positive rate (fpr) is low, and tpr - (1-fpr) is zero or near to zero. Listing 4-3 is the example code to plot a receiver operating characteristic (ROC) plot of tprvs. 1-fpr.

Listing 4-3. Find Optimal Cutoff Point

```
# extract false positive, true positive rate
fpr, tpr, thresholds = metrics.roc_curve(y_train, y_pred_prob)
roc_auc = metrics.auc(fpr, tpr)
print("Area under the ROC curve : %f" % roc_auc)

i = np.arange(len(tpr)) # index for df
roc = pd.DataFrame({'fpr' : pd.Series(fpr, index=i),'tpr' : pd.Series
(tpr, index = i),'1-fpr' : pd.Series(1-fpr, index = i), 'tf' : pd.Series
(tpr - (1-fpr), index = i),'thresholds' : pd.Series(thresholds,
index = i)})
roc.ix[(roc.tf-0).abs().argsort()[:1]]
```

```
# Plot tpr vs 1-fpr
fig, ax = plt.subplots()
plt.plot(roc['tpr'], label='tpr')
plt.plot(roc['1-fpr'], color = 'red', label='1-fpr')
plt.legend(loc='best')
plt.xlabel('1-False Positive Rate')
plt.ylabel('True Positive Rate')
plt.title('Receiver operating characteristic')
plt.show()
#----output----
```

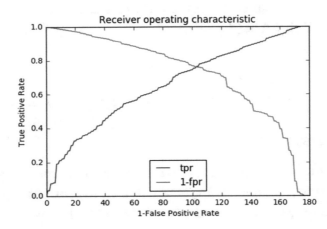

From the chart, the point where tpr crosses 1-fpr is the optimal cutoff point. To simplify finding the optimal probability threshold and enabling reusability, I have made a function to find the optimal probability cutoff point (Listing 4-4).

Listing 4-4. A Function for Finding Optimal Probability Cutoff

```
def Find_Optimal_Cutoff(target, predicted):
    """ Find the optimal probability cutoff point for a classification
        model related to the event rate
    Parameters
    ----------
    target: Matrix with dependent or target data, where rows are
    observations

    predicted: Matrix with predicted data, where rows are observations
```

```
    Returns
    -------
    list type, with optimal cutoff value

    """
    fpr, tpr, threshold = metrics.roc_curve(target, predicted)
    i = np.arange(len(tpr))
    roc = pd.DataFrame({'tf' : pd.Series(tpr-(1-fpr), index=i), 'threshold' :
    pd.Series(threshold, index=i)})
    roc_t = roc.ix[(roc.tf-0).abs().argsort()[:1]]

    return list(roc_t['threshold'])

# Find optimal probability threshold
# Note: probs[:, 1] will have the probability of being a positive label
threshold = Find_Optimal_Cutoff(y_train, probs[:, 1])
print ("Optimal Probability Threshold: ", threshold)

# Applying the threshold to the prediction probability
y_pred_optimal = np.where(y_pred_prob >= threshold, 1, 0)

# Let's compare the accuracy of traditional/normal approach vs optimal cutoff
print ("\nNormal - Accuracy: ", metrics.accuracy_score(y_train, y_pred))
print ("Optimal Cutoff - Accuracy: ", metrics.accuracy_score(y_train,
y_pred_optimal))
print ("\nNormal - Confusion Matrix: \n", metrics.confusion_matrix
(y_train, y_pred))
print ("Optimal - Cutoff Confusion Matrix: \n", metrics.confusion_matrix
(y_train, y_pred_optimal))
#----output----
Optimal Probability Threshold:  [0.36133240553264734]

Normal - Accuracy:  0.767225325885
Optimal Cutoff - Accuracy:  0.761638733706

Normal - Confusion Matrix:
[[303  40]
 [ 85 109]]
```

```
Optimal - Cutoff Confusion Matrix:
[[260  83]
 [ 47 147]]
```

Notice that there is no significant difference in overall accuracy between normal vs. optimal cutoff method; both are 76%. However, there is a 36% increase in true positive rate in the optimal cutoff method (i.e., you are now able to capture 36% more positive cases as positive). Also, the false positive (Type I error) has doubled (i.e., the probability of predicting an individual not having diabetes as positive has increased).

Which Error Is Costly?

Well, there is no one answer to this question! It depends on the domain, the problem that you are trying to address, and the business requirement (Figure 4-1). In our Pima diabetic case, comparatively, type II error might be more damaging than type I error, but it's arguable.

Figure 4-1. *Type I vs. type II error*

Rare Event or Imbalanced Dataset

Providing an equal sample of positive and negative instances to the classification algorithm will result in an optimal result. The dataset that is highly skewed toward one or more classes has proved to be a challenge.

Resampling is a common practice to address the imbalanced dataset issue. Although there are many techniques within resampling, here we'll be learning the three most popular techniques (Figure 4-2):

- *Random undersampling*: Reduce majority class to match minority class count

- *Random oversampling*: Increase minority class by randomly picking samples within minority class till counts of both class match

- *Synthetic Minority Over-Sampling Technique (SMOTE)*: Increase minority class by introducing synthetic examples through connecting all k (default = 5) minority class nearest neighbors, using feature space similarity (Euclidean distance)

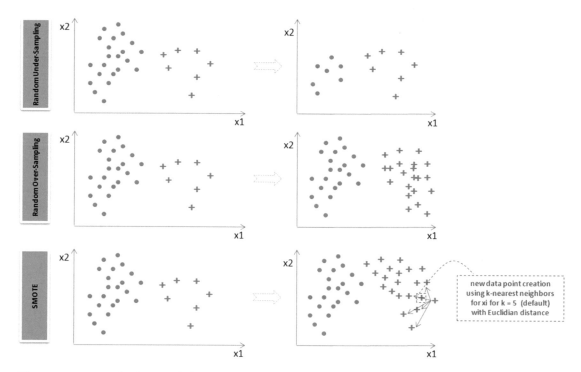

Figure 4-2. *Imbalanced dataset handling techniques*

Let's create a sample imbalanced dataset using the make_classification function of sklearn (Listing 4-5).

Listing 4-5. Rare Event or Imbalanced Data Handling

```
# Load libraries
import matplotlib.pyplot as plt
from sklearn.datasets import make_classification

from imblearn.under_sampling import RandomUnderSampler
from imblearn.over_sampling import RandomOverSampler
from imblearn.over_sampling import SMOTE

# Generate the dataset with 2 features to keep it simple
X, y = make_classification(n_samples=5000, n_features=2, n_informative=2,
                           n_redundant=0, weights=[0.9, 0.1], random_
                           state=2017)

print ("Positive class: ", y.tolist().count(1))
print ("Negative class: ", y.tolist().count(0))
#----output----
Positive class:   514
Negative class:   4486
```

Let's apply the previously described three sampling techniques to the dataset, to balance the dataset and visualize for better understanding.

```
# Apply the random under-sampling
rus = RandomUnderSampler()
X_RUS, y_RUS = rus.fit_sample(X, y)

# Apply the random over-sampling
ros = RandomOverSampler()
X_ROS, y_ROS = ros.fit_sample(X, y)

# Apply regular SMOTE
sm = SMOTE(kind='regular')
X_SMOTE, y_SMOTE = sm.fit_sample(X, y)

# Original vs resampled subplots
plt.figure(figsize=(10, 6))
plt.subplot(2,2,1)
```

```
plt.scatter(X[y==0,0], X[y==0,1], marker='o', color='blue')
plt.scatter(X[y==1,0], X[y==1,1], marker='+', color='red')
plt.xlabel('x1')
plt.ylabel('x2')
plt.title('Original: 1=%s and 0=%s' %(y.tolist().count(1), y.tolist().count(0)))

plt.subplot(2,2,2)
plt.scatter(X_RUS[y_RUS==0,0], X_RUS[y_RUS==0,1], marker='o', color='blue')
plt.scatter(X_RUS[y_RUS==1,0], X_RUS[y_RUS==1,1], marker='+', color='red')
plt.xlabel('x1')
plt.ylabel('y2')
plt.title('Random Under-sampling: 1=%s and 0=%s' %(y_RUS.tolist().count(1),
y_RUS.tolist().count(0)))

plt.subplot(2,2,3)
plt.scatter(X_ROS[y_ROS==0,0], X_ROS[y_ROS==0,1], marker='o', color='blue')
plt.scatter(X_ROS[y_ROS==1,0], X_ROS[y_ROS==1,1], marker='+', color='red')
plt.xlabel('x1')
plt.ylabel('x2')
plt.title('Random over-sampling: 1=%s and 0=%s' %(y_ROS.tolist().count(1),
y_ROS.tolist().count(0)))

plt.subplot(2,2,4)
plt.scatter(X_SMOTE[y_SMOTE==0,0], X_SMOTE[y_SMOTE==0,1], marker='o',
color='blue')
plt.scatter(X_SMOTE[y_SMOTE==1,0], X_SMOTE[y_SMOTE==1,1], marker='+',
color='red')
plt.xlabel('x1')
plt.ylabel('y2')
plt.title('SMOTE: 1=%s and 0=%s' %(y_SMOTE.tolist().count(1), y_SMOTE.
tolist().count(0)))

plt.tight_layout()
plt.show()
#----output----
```

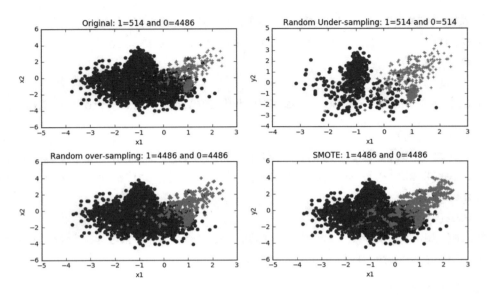

Caution   Keep in mind that random undersampling raises the opportunity for loss of information or concepts, as we are reducing the majority class, and that random oversampling & SMOTE can lead to an overfitting issue due to multiple related instances.

Which Resampling Technique Is the Best?

Well, yet again there is no one answer to this question! Let's try a quick classification model on the preceding three resampled data and compare the accuracy. We'll use the AUC metric, as this is one of the best representations of model performance (Listing 4-6).

Listing 4-6. Build Models on Various Resampling Methods and Evaluate Performance

```
from sklearn import tree
from sklearn import metrics
from sklearn.cross_ model_selection import train_test_split

X_RUS_train, X_RUS_test, y_RUS_train, y_RUS_test = train_test_split(X_RUS,
y_RUS, test_size=0.3, random_state=2017)
X_ROS_train, X_ROS_test, y_ROS_train, y_ROS_test = train_test_split(X_ROS,
y_ROS, test_size=0.3, random_state=2017)
```

```
X_SMOTE_train, X_SMOTE_test, y_SMOTE_train, y_SMOTE_test = train_test_
split(X_SMOTE, y_SMOTE, test_size=0.3, random_state=2017)

# build a decision tree classifier
clf = tree.DecisionTreeClassifier(random_state=2017)
clf_rus = clf.fit(X_RUS_train, y_RUS_train)
clf_ros = clf.fit(X_ROS_train, y_ROS_train)
clf_smote = clf.fit(X_SMOTE_train, y_SMOTE_train)

# evaluate model performance
print ("\nRUS - Train AUC : ",metrics.roc_auc_score(y_RUS_train, clf.
predict(X_RUS_train)))
print ("RUS - Test AUC : ",metrics.roc_auc_score(y_RUS_test, clf.predict
(X_RUS_test)))
print ("ROS - Train AUC : ",metrics.roc_auc_score(y_ROS_train, clf.
predict(X_ROS_train)))
print ("ROS - Test AUC : ",metrics.roc_auc_score(y_ROS_test, clf.predict
(X_ROS_test)))
print ("\nSMOTE - Train AUC : ",metrics.roc_auc_score(y_SMOTE_train, clf.
predict(X_SMOTE_train)))
print ("SMOTE - Test AUC : ",metrics.roc_auc_score(y_SMOTE_test, clf.
predict(X_SMOTE_test)))
#----output----

RUS - Train AUC :  0.988945248974
RUS - Test AUC :  0.983964646465
ROS - Train AUC :  0.985666951094
ROS - Test AUC :  0.986630288452

SMOTE - Train AUC :  1.0
SMOTE - Test AUC :  0.956132695918
```

Here random oversampling is performing better on both train and test sets. As a best practice, in real-world use cases it is recommended to look at other metrics (such as precision, recall, confusion matrix) and apply business context or domain knowledge to assess the true performance of the model.

Bias and Variance

A fundamental problem with supervised learning is the bias–variance tradeoff. Ideally, a model should have two key characteristics:

1. It should be sensitive enough to accurately capture the key patterns in the training dataset.

2. It should be generalized enough to work well on any unseen datasets.

Unfortunately, while trying to achieve the aforementioned first point, there is an ample chance of overfitting to noisy or unrepresentative training data points, leading to a failure of generalizing the model. On the other hand, trying to generalize a model may result in failing to capture important regularities (Figure 4-3).

Bias

If model accuracy is low on the training dataset as well as the test dataset, the model is said to be underfitting or has a high bias. This means the model is not fitting the training dataset points well in regression or the decision boundary is not separating the classes well in classification. Two key reasons for bias are 1) not including the right features and 2) not picking the correct order of polynomial degree for model fitting.

To solve the underfitting issue or to reduced bias, try including more meaningful features and try to increase the model complexity by trying higher order polynomial fittings.

Variance

If a model is giving high accuracy on the training dataset, but on the test dataset the accuracy drops drastically, then the model is said to be overfitting or has high variance. The key reason for overfitting is using a higher order polynomial degree (may not be required), which will fit the decision boundary tool well to all data points including the noise of the train dataset instead of the underlying relationship. This will lead to a high accuracy (actual vs. predicted) in the train dataset and when applied to the test dataset, the prediction error will be high.

To solve the overfitting issue:

- Try to reduce the number of features, that is, keep only the meaningful features or try regularization methods that will keep all the features but reduce the magnitude of the feature parameter.

- Dimension reduction can eliminate noisy features, in turn reducing the model variance.

- Bringing more data points to make training dataset large will also reduce variance.

- Choosing the right model parameters can help to reduce the bias and variance, for example.

  - Using right regularization parameters can decrease variance in regression-based models.

  - For a decision tree, reducing the depth of the decision tree will reduce the variance.

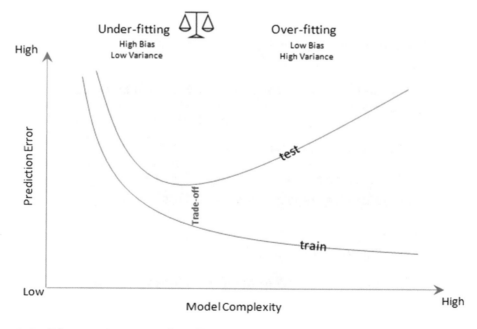

Figure 4-3. *Bias-variance trade-off*

K-Fold Cross Validation

K-fold cross-validation splits the training dataset into k folds without replacement—any given data point will only be part of one of the subsets, where k-1 folds are used for the model training and one fold is used for testing. The procedure is repeated k times so that we obtain k models and performance estimates (Figure 4-4).

We then calculate the average performance of the models based on the individual folds, to obtain a performance estimate that is less sensitive to the subpartitioning of the training data compared with the holdout or single fold method.

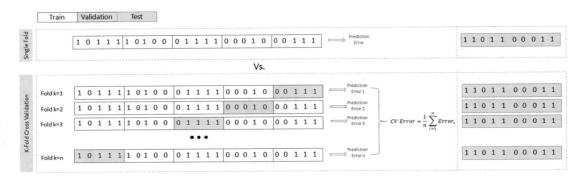

Figure 4-4. *K-fold cross-validation*

Listing 4-7 shows an example code to use k-fold cross-validation of sklearn to build a classification model.

Listing 4-7. K-fold Cross-Validation

```
from sklearn.model_selection import cross_val_score
from sklearn.preprocessing import StandardScaler

# read the data in
df = pd.read_csv("Data/Diabetes.csv")

X = df.ix[:,:8].values      # independent variables
y = df['class'].values      # dependent variables
```

```
# Normalize Data
sc = StandardScaler()
sc.fit(X)
X = sc.transform(X)

# evaluate the model by splitting into train and test sets
X_train, X_test, y_train, y_test = train_test_split(X, y, test_size=0.3,
random_state=2017)

# build a decision tree classifier
clf = tree.DecisionTreeClassifier(random_state=2017)

# evaluate the model using 10-fold cross-validation
train_scores = cross_val_score(clf, X_train, y_train, scoring='accuracy', cv=5)
test_scores = cross_val_score(clf, X_test, y_test, scoring='accuracy', cv=5)
print ("Train Fold AUC Scores: ", train_scores)
print ("Train CV AUC Score: ", train_scores.mean())

print ("\nTest Fold AUC Scores: ", test_scores)
print ("Test CV AUC Score: ", test_scores.mean())
#---output----
Train Fold AUC Scores:  [0.80555556 0.73148148 0.81308411 0.76635514
0.71028037]
Train CV AUC Score:  0.7653513326410523

Test Fold AUC Scores:  [0.80851064 0.78723404 0.78723404 0.77777778 0.8    ]
Test CV AUC Score:  0.7921513002364066
```

Stratified K-fold Cross-Validation

Extended cross-validation is the stratified k-fold cross-validation, where the class proportions are preserved in each fold, leading to better bias and variance estimates (Listing 4-8 and 4-9).

Listing 4-8. Stratified K-fold Cross-Validation

```python
from sklearn import model_selection

kfold = model_selection.StratifiedKFold(n_splits=5, random_state=2019)

train_scores = []
test_scores = []
k = 0
for (train, test) in kfold.split(X_train, y_train):
    clf.fit(X_train[train], y_train[train])
    train_score = clf.score(X_train[train], y_train[train])
    train_scores.append(train_score)
    # score for test set
    test_score = clf.score(X_train[test], y_train[test])
    test_scores.append(test_score)

    k += 1
    print('Fold: %s, Class dist.: %s, Train Acc: %.3f, Test Acc: %.3f'
            % (k, np.bincount(y_train[train]), train_score, test_score))
print('\nTrain CV accuracy: %.3f' % (np.mean(train_scores)))
print('Test CV accuracy: %.3f' % (np.mean(test_scores)))
#----output----
Fold: 1, Class dist.: [277 152], Train Acc: 0.758, Test Acc: 0.806
Fold: 2, Class dist.: [277 152], Train Acc: 0.779, Test Acc: 0.731
Fold: 3, Class dist.: [278 152], Train Acc: 0.767, Test Acc: 0.813
Fold: 4, Class dist.: [278 152], Train Acc: 0.781, Test Acc: 0.766
Fold: 5, Class dist.: [278 152], Train Acc: 0.781, Test Acc: 0.710

Train CV accuracy: 0.773
Test CV accuracy: 0.765
```

Listing 4-9. Plotting the ROC Curve for Stratified K-fold Cross-Validation

```python
from sklearn.metrics import roc_curve, auc
from itertools import cycle
from scipy import interp

kfold = model_selection.StratifiedKFold(n_splits=5, random_state=2019)
```

```
mean_tpr = 0.0
mean_fpr = np.linspace(0, 1, 100)

colors = cycle(['cyan', 'indigo', 'seagreen', 'yellow', 'blue', 'darkorange'])
lw = 2

i = 0
for (train, test), color in zip(kfold.split(X, y), colors):
    probas_ = clf.fit(X[train], y[train]).predict_proba(X[test])
    # Compute ROC curve and area the curve
    fpr, tpr, thresholds = roc_curve(y[test], probas_[:, 1])
    mean_tpr += interp(mean_fpr, fpr, tpr)
    mean_tpr[0] = 0.0
    roc_auc = auc(fpr, tpr)
    plt.plot(fpr, tpr, lw=lw, color=color,
             label='ROC fold %d (area = %0.2f)' % (i, roc_auc))

    i += 1
plt.plot([0, 1], [0, 1], linestyle='--', lw=lw, color='k',
         label='Luck')

mean_tpr /= kfold.get_n_splits(X, y)
mean_tpr[-1] = 1.0
mean_auc = auc(mean_fpr, mean_tpr)
plt.plot(mean_fpr, mean_tpr, color='g', linestyle='--',
         label='Mean ROC (area = %0.2f)' % mean_auc, lw=lw)

plt.xlim([-0.05, 1.05])
plt.ylim([-0.05, 1.05])
plt.xlabel('False Positive Rate')
plt.ylabel('True Positive Rate')
plt.title('Receiver operating characteristic example')
plt.legend(loc="lower right")
plt.show()
#----Output----
```

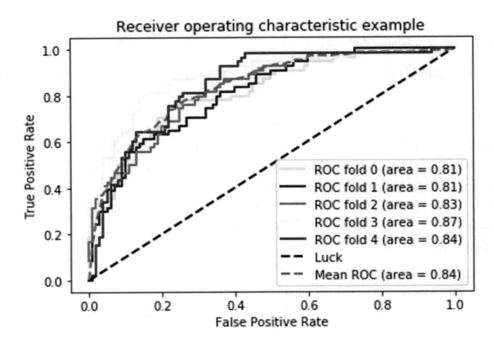

Ensemble Methods

Ensemble methods enable combining multiple model scores into a single score to create a robust generalized model.

At a high level, there are two types of ensemble methods:

1. Combine multiple models of similar type.

    - Bagging (bootstrap aggregation)

    - Boosting

2. Combine multiple models of various types.

    - Vote classification

    - Blending or stacking

Bagging

Bootstrap aggregation (also known as bagging) was proposed by Leo Breiman in 1994; it is a model aggregation technique to reduce model variance. The training data is split into multiple samples with a replacement called bootstrap samples. Bootstrap sample size will be the same as the original sample size, with 3/4 of the original values and replacement resulting in repetition of values (Figure 4-5).

Original Sample	1	2	3	4	5	6	7	8	9	10
Bootstrap Sample 1	3	1	6	5	10	7	7	9	2	3
Bootstrap Sample 2	3	10	9	3	9	8	7	10	2	10
Bootstrap Sample 3	3	5	4	10	7	7	6	2	6	9
Bootstrap Sample 4	10	10	10	3	6	2	5	4	7	9

Bootstrap sample size will be same as original sample size, with ¾ of the original values + replacement result in repetition of values

***Figure 4-5.** Bootstrapping*

Independent models on each of the bootstrap samples are built, and the average of the predictions for regression or majority vote for classification is used to create the final model.

Figure 4-6 shows the bagging process flow. If N is the number of bootstrap samples created out of the original training set, for i = 1 to N, train a base ML model C_i.

$$C_{final} = \text{aggregate max of y} \sum_i I(C_i = y)$$

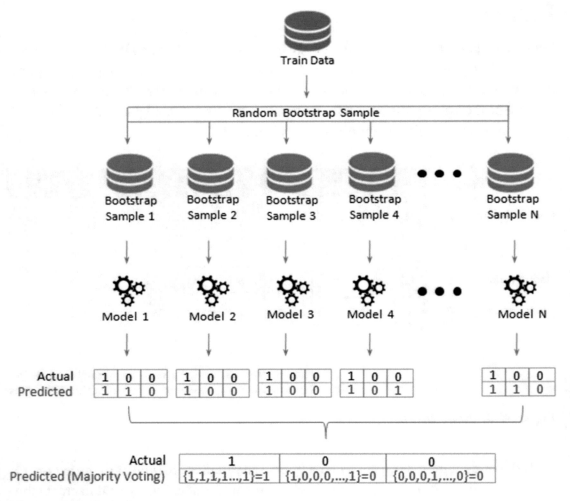

Figure 4-6. *Bagging*

Let's compare the performance of a stand-alone decision tree model and a bagging decision tree model of 100 trees (Listing 4-10).

Listing 4-10. Stand-Alone Decision Tree vs. Bagging

```
# Bagged Decision Trees for Classification
from sklearn.ensemble import BaggingClassifier
from sklearn.tree import DecisionTreeClassifier
```

```
# read the data in
df = pd.read_csv("Data/Diabetes.csv")

X = df.ix[:,:8].values      # independent variables
y = df['class'].values      # dependent variables

#Normalize
X = StandardScaler().fit_transform(X)

# evaluate the model by splitting into train and test sets
X_train, X_test, y_train, y_test = train_test_split(X, y, test_size=0.2,
random_state=2019)

kfold = model_selection.StratifiedKFold(n_splits=5, random_state=2019)
num_trees = 100

# Decision Tree with 5 fold cross validation
clf_DT = DecisionTreeClassifier(random_state=2019).fit(X_train,y_train)
results = model_selection.cross_val_score(clf_DT, X_train,y_train, cv=kfold)
print ("Decision Tree (stand alone) - Train : ", results.mean())
print ("Decision Tree (stand alone) - Test : ", metrics.accuracy_score(clf_
DT.predict(X_test), y_test))

# Using Bagging Lets build 100 decision tree models and average/majority
vote prediction
clf_DT_Bag = BaggingClassifier(base_estimator=clf_DT, n_estimators=num_
trees, random_state=2019).fit(X_train,y_train)
results = model_selection.cross_val_score(clf_DT_Bag, X_train, y_train,
cv=kfold)
print ("\nDecision Tree (Bagging) - Train : ", results.mean())
print ("Decision Tree (Bagging) - Test : ", metrics.accuracy_score(clf_DT_
Bag.predict(X_test), y_test))
#----output----
Decision Tree (stand alone) - Train :  0.6742199894235854
Decision Tree (stand alone) - Test :  0.6428571428571429

Decision Tree (Bagging) - Train :  0.7460074034902167
Decision Tree (Bagging) - Test :  0.8051948051948052
```

Feature Importance

The decision tree model has an attribute to show important features, which are based on the Gini or entropy information gain (Listing 4-11).

Listing 4-11. Decision Tree Feature Importance Function

```
feature_importance = clf_DT.feature_importances_
# make importances relative to max importance
feature_importance = 100.0 * (feature_importance / feature_importance.
max())
sorted_idx = np.argsort(feature_importance)
pos = np.arange(sorted_idx.shape[0]) + .5
plt.subplot(1, 2, 2)
plt.barh(pos, feature_importance[sorted_idx], align='center')
plt.yticks(pos, df.columns[sorted_idx])
plt.xlabel('Relative Importance')
plt.title('Variable Importance')
plt.show()
#----output----
```

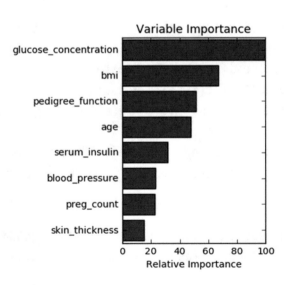

RandomForest

A subset of observations and a subset of variables are randomly picked to build multiple independent tree-based models. The trees are more uncorrelated, as only a subset of variables is used during the split of the tree rather than greedily choosing the best split point in the construction of the tree (Listing 4-12).

Listing 4-12. RandomForest Classifier

```
from sklearn.ensemble import RandomForestClassifier
num_trees = 100

kfold = model_selection.StratifiedKFold(n_splits=5, random_state=2019)

clf_RF = RandomForestClassifier(n_estimators=num_trees).fit(X_train,
y_train)
results = model_selection.cross_val_score(clf_RF, X_train, y_train, cv=kfold)

print ("\nRandom Forest (Bagging) - Train : ", results.mean())
print ("Random Forest (Bagging) - Test : ", metrics.accuracy_score(clf_
RF.predict(X_test), y_test))
#----output----
Random Forest - Train :  0.7379693283976732
Random Forest - Test :  0.8051948051948052
```

Extremely Randomized Trees (ExtraTree)

This algorithm is an effort to introduce more randomness to the bagging process. Tree splits are chosen completely at random from the range of values in the sample at each split, which allows reducing the variance of the model further but at the cost of a slight increase in bias (Listing 4-13).

Listing 4-13. Extremely Randomized Trees (ExtraTree)

```
from sklearn.ensemble import ExtraTreesClassifier
num_trees = 100

kfold = model_selection.StratifiedKFold(n_splits=5, random_state=2019)
```

```
clf_ET = ExtraTreesClassifier(n_estimators=num_trees).fit(X_train, y_train)
results = cross_validation.cross_val_score(clf_ET, X_train, y_train, cv=kfold)

print ("\nExtraTree - Train : ", results.mean())
print ("ExtraTree - Test : ", metrics.accuracy_score(clf_ET.predict(X_test),
y_test))
#----output----
ExtraTree - Train :   0.7410893707033315
ExtraTree - Test :   0.7987012987012987
```

How Does the Decision Boundary Look?

Let's perform PCA and consider only the first two principal components for easy plotting. The model building code would remain the same as before except that after normalization and before splitting the data to train and test, we will need to add the following code.

Once we have run the model successfully, we can use the following code to draw decision boundaries for stand-alone vs. different bagging models.

Listing 4-14. Plot the Decision Boudaries

```
from sklearn.decomposition import PCA
from matplotlib.colors import ListedColormap
# PCA
X = PCA(n_components=2).fit_transform(X)

def plot_decision_regions(X, y, classifier):

    h = .02  # step size in the mesh
    # setup marker generator and color map
    markers = ('s', 'x', 'o', '^', 'v')
    colors = ('red', 'blue', 'lightgreen', 'gray', 'cyan')
    cmap = ListedColormap(colors[:len(np.unique(y))])

    # plot the decision surface
    x1_min, x1_max = X[:, 0].min() - 1, X[:, 0].max() + 1
    x2_min, x2_max = X[:, 1].min() - 1, X[:, 1].max() + 1
```

```python
    xx1, xx2 = np.meshgrid(np.arange(x1_min, x1_max, h),
                           np.arange(x2_min, x2_max, h))
    Z = classifier.predict(np.array([xx1.ravel(), xx2.ravel()]).T)
    Z = Z.reshape(xx1.shape)
    plt.contourf(xx1, xx2, Z, alpha=0.4, cmap=cmap)
    plt.xlim(xx1.min(), xx1.max())
    plt.ylim(xx2.min(), xx2.max())

    for idx, cl in enumerate(np.unique(y)):
        plt.scatter(x=X[y == cl, 0], y=X[y == cl, 1],
                    alpha=0.8, c=colors[idx],
                    marker=markers[idx], label=cl)

# Plot the decision boundary
plt.figure(figsize=(10,6))
plt.subplot(221)
plot_decision_regions(X, y, clf_DT)
plt.title('Decision Tree (Stand alone)')
plt.xlabel('PCA1')
plt.ylabel('PCA2')

plt.subplot(222)
plot_decision_regions(X, y, clf_DT_Bag)
plt.title('Decision Tree (Bagging - 100 trees)')
plt.xlabel('PCA1')
plt.ylabel('PCA2')
plt.legend(loc='best')

plt.subplot(223)
plot_decision_regions(X, y, clf_RF)
plt.title('RandomForest Tree (100 trees)')
plt.xlabel('PCA1')
plt.ylabel('PCA2')
plt.legend(loc='best')

plt.subplot(224)
plot_decision_regions(X, y, clf_ET)
plt.title('Extream Random Tree (100 trees)')
```

```
plt.xlabel('PCA1')
plt.ylabel('PCA2')
plt.legend(loc='best')
plt.tight_layout()

#----output----

Decision Tree (stand alone) - Train :  0.5781332628239026
Decision Tree (stand alone) - Test :  0.6688311688311688

Decision Tree (Bagging) - Train :  0.6319936541512428
Decision Tree (Bagging) - Test :  0.7467532467532467

Random Forest - Train :  0.6418297197250132
Random Forest  - Test :  0.7662337662337663

ExtraTree - Train :  0.6205446853516658
ExtraTree - Test :  0.7402597402597403
```

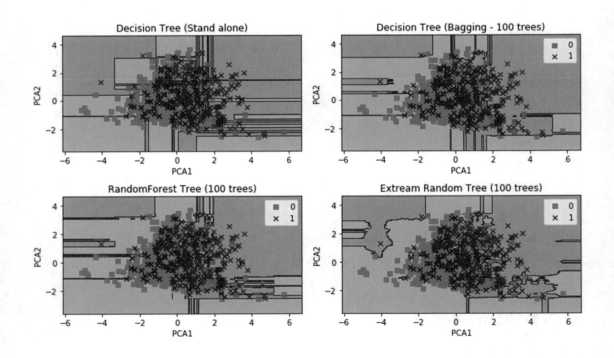

Bagging—Essential Tuning Parameters

Let's look at the key tuning parameters to get better model results.

- *n_estimators:* This is the number of trees—the larger the better. Note that beyond a certain point, the results will not improve significantly.

- *max_features:* This is the random subset of features to be used for splitting a node—the lower the better to reduce variance (but increases bias). Ideally, for a regression problem it should be equal to n_features (total number of features) and for classification, square root of n_features.

- *n_jobs:* Number of cores to be used for parallel construction of tress. If set to -1, all available cores in the system are used, or you can specify the number.

Boosting

Freud and Schapire in 1995 introduced the concept of boosting with the well-known AdaBoost algorithm (adaptive boosting). The core concept of boosting is that rather than an independent individual hypothesis, combining hypotheses in a sequential order increases the accuracy. Essentially, boosting algorithms convert the weak learners into strong learners. Boosting algorithms are well designed to address bias problems (Figure 4-7).

At a high level the AdaBoosting process can be divided into three steps:

- Assign uniform weights for all data points $W_0(x) = 1 / N$, where N is the total number of training data points.

- At each iteration, fit a classifier $y_m(x_n)$ to the training data and update weights to minimize the weighted error function.

  - The weight is calculated as $W_n^{(m+1)} = W_n^{(m)} \exp\left\{ \propto_m y_m \left(x_n \right) \neq t_n \right\}$.

- The hypothesis weight or the loss function is given by $\propto_m = \frac{1}{2}\log\left\{\frac{1-\epsilon_m}{\epsilon_m}\right\}$, and the term rate

  is given by $\epsilon_m = \frac{\sum_{n=1}^{N} W_n^{(m)}\, I\left(y_m(x_n) \neq t_n\right)}{\sum_{n=1}^{N} W_n^{(m)}}$, where

  $\left(y_m(x_n) \neq t_n\right)$ has values $\frac{0}{1}$ i.e., 0 if (x_n) correctly classified else 1

- The final model is given by $Y_m = \text{sign}\left(\sum_{m=1}^{M} \propto_m y_m(x)\right)$

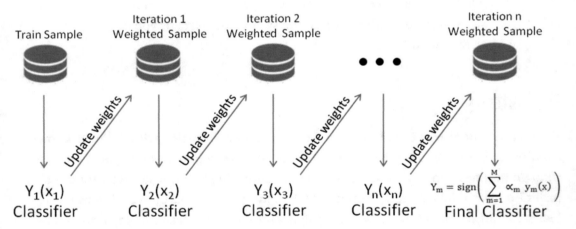

Figure 4-7. *AdaBoosting*

Example Illustration for AdaBoost

Let's consider training data with two class labels of ten data points. Assume, initially, all the data points will have equal weights given by 1/10, as shown in Figure 4-8.

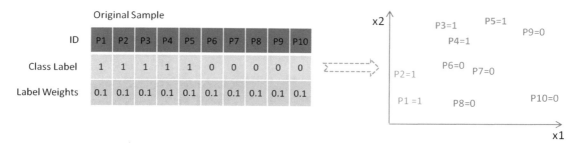

Figure 4-8. *Sample data set with ten data points*

Boosting Iteration 1

Notice in Figure 4-9 that three points of the positive class are misclassified by the first simple classification model, so they will be assigned higher weights. Error term and loss function (learning rate) are calculated as 0.30 and 0.42, respectively. The data points P3, P4, and P5 will get higher weight (0.15) due to misclassification, whereas other data points will retain the original weight (0.1).

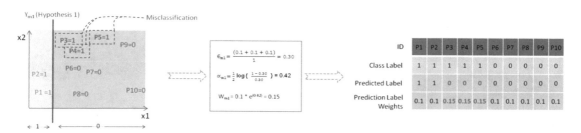

Figure 4-9. Y_{m1} *is the first classification or hypothesis*

Boosting Iteration 2

Let's fit another classification model as shown in Figure 4-10, and notice that three data points (P6, P7, and P8) of the negative class are misclassified. Hence, these will be assigned higher weights of 0.17 as calculated, whereas the remaining data points' weights will remain the same because they are correctly classified.

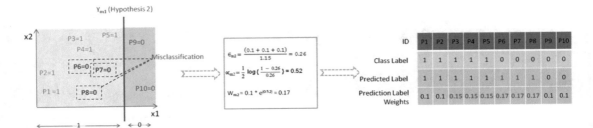

Figure 4-10. *Y_{m2} is the second classification or hypothesis*

Boosting Iteration 3

The third classification model has misclassified a total of three data points: two positive classes, P1 and P2; and one negative class, P9. So these misclassified data points will be assigned a new higher weight of 0.19 as calculated, and the remaining data points will retain their earlier weights (Figure 4-11).

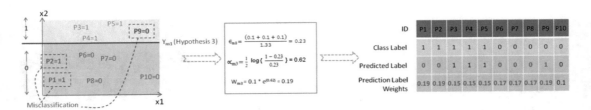

Figure 4-11. *Y_{m3} is the third classification or hypothesis*

Final Model

Now, as per the AdaBoost algorithm, let's combine the weak classification models as shown in Figure 4-12. Notice that the final combined model will have a minimum error term and maximum learning rate, leading to a higher degree of accuracy.

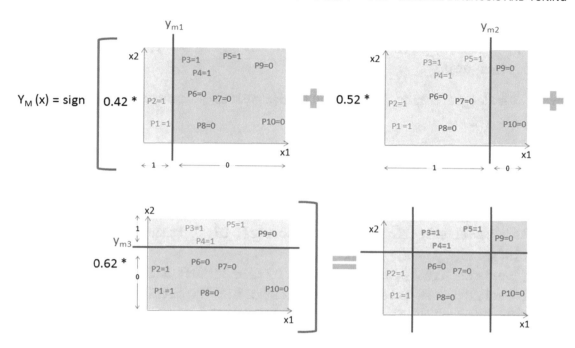

Figure 4-12. *AdaBoost algorithm to combine weak classifiers*

Let's pick weak predictors from the Pima diabetic dataset and compare the performance of a stand-alone decision tree model vs. AdaBoost with 100 boosting rounds on the decision tree model (Listing 4-15).

Listing 4-15. Stand-Alone Decision Tree vs. AdaBoost

```
# Bagged Decision Trees for Classification
from sklearn.ensemble import AdaBoostClassifier
from sklearn.tree import DecisionTreeClassifier

# read the data in
df = pd.read_csv("Data/Diabetes.csv")

# Let's use some week features to build the tree
X = df[['age','serum_insulin']]    # independent variables
y = df['class'].values             # dependent variables

#Normalize
X = StandardScaler().fit_transform(X)
```

```
# evaluate the model by splitting into train and test sets
X_train, X_test, y_train, y_test = train_test_split(X, y, test_size=0.2,
random_state=2019)

kfold = model_selection.StratifiedKFold(n_splits=5, random_state=2019)
num_trees = 100

# Dection Tree with 5 fold cross validation
# lets restrict max_depth to 1 to have more impure leaves
clf_DT = DecisionTreeClassifier(max_depth=1, random_state=2019).
fit(X_train,y_train)
results = model_selection.cross_val_score(clf_DT, X_train,y_train,
cv=kfold.split(X_train, y_train))
print("Decision Tree (stand alone) - CV Train : %.2f" % results.mean())
print("Decision Tree (stand alone) - Test : %.2f" % metrics.accuracy_
score(clf_DT.predict(X_train), y_train))
print("Decision Tree (stand alone) - Test : %.2f" % metrics.accuracy_
score(clf_DT.predict(X_test), y_test))

# Using Adaptive Boosting of 100 iteration
clf_DT_Boost = AdaBoostClassifier(base_estimator=clf_DT, n_estimators=num_
trees, learning_rate=0.1, random_state=2019).fit(X_train,y_train)
results = model_selection.cross_val_score(clf_DT_Boost, X_train, y_train,
cv=kfold.split(X_train, y_train))
print("\nDecision Tree (AdaBoosting) - CV Train : %.2f" % results.mean())
print("Decision Tree (AdaBoosting) - Train : %.2f" % metrics.accuracy_
score(clf_DT_Boost.predict(X_train), y_train))
print("Decision Tree (AdaBoosting) - Test : %.2f" % metrics.accuracy_
score(clf_DT_Boost.predict(X_test), y_test))
#----output----
Decision Tree (stand alone) - CV Train : 0.64
Decision Tree (stand alone) - Test : 0.64
Decision Tree (stand alone) - Test : 0.70

Decision Tree (AdaBoosting) - CV Train : 0.68
Decision Tree (AdaBoosting) - Train : 0.71
Decision Tree (AdaBoosting) - Test : 0.79
```

Notice that in this case, the AdaBoost algorithm has given an average rise of 9% inaccuracy score between train/test dataset compared with the stand-alone decision tree model.

Gradient Boosting

Due to the stagewise additivity, the loss function can be represented in a form suitable for optimization. This gave birth to a class of generalized boosting algorithms known as generalized boosting machine (GBM). Gradient boosting is an example implementation of GBM and it can work with different loss functions such as regression, classification, risk modeling, etc. As the name suggests, it is a boosting algorithm that identifies shortcomings of a weak learner by gradients (AdaBoost uses high-weight data points), hence the name gradient boosting.

- Iteratively fit a classifier $y_m(x_n)$ to the training data. The initial model will be with a constant value $y_0(x) = \arg\min\delta\sum_{i=1}^{n}L(y_m, \delta)$.

- Calculate the loss (i.e., the predicted value vs. actual value) for each model fit iteration $g_m(x)$ or compute the negative gradient and use it to fit a new base learner function $h_m(x)$, and find the best gradient descent step-size $\delta_m = \arg\min\delta\sum_{i=1}^{n}L\big(y_m, y_{m-1}(x) + \delta h_m(x)\big)$.

- Update the function estimate $y_m(x) = y_{m-1}(x) + \delta h_m(x)$ and output $y_m(x)$.

Listing 4-16 shows an example code implementation of a gradient boosting classifier.

Listing 4-16. Gradient Boosting Classifier

```
from sklearn.ensemble import GradientBoostingClassifier

# Using Gradient Boosting of 100 iterations
clf_GBT = GradientBoostingClassifier(n_estimators=num_trees, learning_
rate=0.1, random_state=2019).fit(X_train, y_train)
results = model_selection.cross_val_score(clf_GBT, X_train, y_train,
cv=kfold)
```

```
print ("\nGradient Boosting - CV Train : %.2f" % results.mean())
print ("Gradient Boosting - Train : %.2f" % metrics.accuracy_score(clf_GBT.
predict(X_train), y_train))
print ("Gradient Boosting - Test : %.2f" % metrics.accuracy_score(clf_GBT.
predict(X_test), y_test))
#----output----
Gradient Boosting - CV Train : 0.66
Gradient Boosting - Train : 0.79
Gradient Boosting - Test : 0.75
```

Let's look at the digit classification to illustrate how the model performance improves with each iteration.

```
from sklearn.ensemble import GradientBoostingClassifier

df= pd.read_csv('Data/digit.csv')

X = df.iloc[:,1:17].values
y = df['lettr'].values

# evaluate the model by splitting into train and test sets
X_train, X_test, y_train, y_test = train_test_split(X, y, test_size=0.2,
random_state=2019)

kfold = model_selection.StratifiedKFold(n_splits=5, random_state=2019)
num_trees = 10

clf_GBT = GradientBoostingClassifier(n_estimators=num_trees, learning_
rate=0.1, random_state=2019).fit(X_train, y_train)
results = model_selection.cross_val_score(clf_GBT, X_train, y_train, cv=kfold)

print ("\nGradient Boosting - Train : %.2f" % metrics.accuracy_score
(clf_GBT.predict(X_train), y_train))
print ("Gradient Boosting - Test : %.2f" % metrics.accuracy_score
(clf_GBT.predict(X_test), y_test))

# Let's predict for the letter 'T' and understand how the prediction
accuracy changes in each boosting iteration
X_valid= (2,8,3,5,1,8,13,0,6,6,10,8,0,8,0,8)
print ("Predicted letter: ", clf_GBT.predict([X_valid]))
```

```
# Staged prediction will give the predicted probability for each boosting
iteration
stage_preds = list(clf_GBT.staged_predict_proba([X_valid]))
final_preds = clf_GBT.predict_proba([X_valid])

# Plot
x = range(1,27)
label = np.unique(df['lettr'])

plt.figure(figsize=(10,3))
plt.subplot(131)
plt.bar(x, stage_preds[0][0], align='center')
plt.xticks(x, label)
plt.xlabel('Label')
plt.ylabel('Prediction Probability')
plt.title('Round One')
plt.autoscale()

plt.subplot(132)
plt.bar(x, stage_preds[5][0],align='center')
plt.xticks(x, label)
plt.xlabel('Label')
plt.ylabel('Prediction Probability')
plt.title('Round Five')
plt.autoscale()

plt.subplot(133)
plt.bar(x, stage_preds[9][0],align='center')
plt.xticks(x, label)
plt.autoscale()
plt.xlabel('Label')
plt.ylabel('Prediction Probability')
plt.title('Round Ten')

plt.tight_layout()
plt.show()
#----output----
Gradient Boosting - Train :  0.75
```

```
Gradient Boosting - Test :   0.72
Predicted letter: 'T'
```

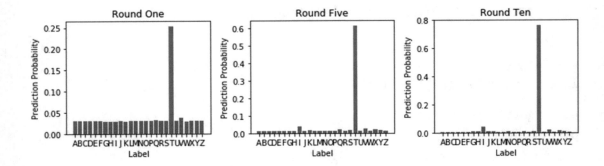

Gradient boosting corrects the erroneous boosting iteration's negative impact in subsequent iterations. Notice that in the first iteration the predicted probability for letter "T" is 0.25 and it gradually increased to 0.76 by the tenth iteration, whereas the probability percentage for other letters have decreased over each round.

Boosting—Essential Tuning Parameters

Model complexity and overfitting can be controlled by using correct values for two categories of parameters:

1. Tree structure

    n_estimators: This is the number of weak learners to be built.

    max_depth: This is the maximum depth of the individual estimators. The best value depends on the interaction of the input variables.

    min_samples_leaf: This will be helpful to ensure a sufficient number of samples results in leaf.

    subsample: This is the fraction of the sample to be used for fitting individual models (default=1). Typically .8 (80%) is used to introduce a random selection of samples, which in turn increases the robustness against overfitting.

2. Regularization parameter

learning_rate: This controls the magnitude of change in estimators. The lower learning rate is better, which requires higher n_estimators (that is the trade-off).

Xgboost (eXtreme Gradient Boosting)

In March 2014, Tianqui Chen built xgboost in C++ as part of the distributed (deep) ML community, and it has an interface for Python. It is an extended, more regularized version of a gradient boosting algorithm. This is one of the most well-performing, large-scale, scalable ML algorithms, which has been playing a major role in winning solutions in Kaggle (forum for predictive modeling and analytics competition) data science competition.

XGBoost objective function $\text{obj}(\Theta) = \sum_i^n l(y_i - \hat{y}_i) + \sum_{k=1}^{K} \Omega(f_k)$

Regularization term is given by

$$\Omega(f) = \gamma T + \frac{1}{2} \lambda \sum_{j=1}^{T} W_j^2$$

Controls the overall   +   Scores of the overall
number of leaves created   number of leaves created

Gradient descent technique is used for optimizing the objective function, and more mathematics about the algorithms can be found at the site `http://xgboost.readthedocs.io/en/latest/`.

Some of the key advantages of the xgboost algorithm are

- It implements parallel processing.

- It has a built-in standard to handle missing values, which means the user can specify a particular value different than other observations (such as -1 or -999) and pass it as a parameter.

- It will split the tree up to maximum depth, unlike gradient boosting where it stops splitting the node on encountering a negative loss in the split.

XGboost has a bundle of parameters, and at a high level we can group them into three categories. Let's look at the most important within these categories.

1. General parameters

    a. *nthread*: Number of parallel threads; if not given a value all cores will be used.

    b. *Booster*: This is the type of model to be run, with gbtree (tree-based model) being the default. "gblinear" to be used for linear models

2. Boosting parameters

    a. *eta*: This is the learning rate or step size shrinkage to prevent overfitting; default is 0.3 and it can range between 0 and 1.

    b. *max_depth*: Maximum depth of tree, with the default being 6

    c. *min_child_weight*: The minimum sum of weights of all observations required in a child. Start with the 1/square root of event rate.

    d. *colsample_bytree*: A fraction of columns to be randomly sampled for each tree, with a default value of 1.

    e. *Subsample*: A fraction of observations to be randomly sampled for each tree, with a default value of 1. Lowering this value makes the algorithm conservative to avoid overfitting.

    f. *lambda*: L2 regularization term on weights, with a default value of 1

    g. *alpha*: L1 regularization term on weight

3. Task parameters

    a. *Objective*: This defines the loss function to be minimized, with default value "reg: linear." For binary classification it should be "binary: logistic" and for multiclass "multi:softprob" to get the probability value and "multi: softmax" to get predicted class. For multiclass, num_class (number of unique classes) is to be specified.

    b. *eval_metric*: Metric to be used for validating model performance

sklearn has a wrapper for xgboost (XGBClassifier). Let's continue with the diabetics data set and build a model using the weak learner (Listing 4-17).

Listing 4-17. xgboost Classifier Using sklearn Wrapper

```python
import xgboost as xgb
from xgboost.sklearn import XGBClassifier

# read the data in
df = pd.read_csv("Data/Diabetes.csv")

predictors = ['age','serum_insulin']
target = 'class'

# Most common preprocessing step include label encoding and missing value
treatment
from sklearn import preprocessing
for f in df.columns:
    if df[f].dtype=='object':
        lbl = preprocessing.LabelEncoder()
        lbl.fit(list(df[f].values))
        df[f] = lbl.transform(list(df[f].values))

df.fillna((-999), inplace=True)

# Let's use some week features to build the tree
X = df[['age','serum_insulin']] # independent variables
y = df['class'].values          # dependent variables

#Normalize
X = StandardScaler().fit_transform(X)

# evaluate the model by splitting into train and test sets
X_train, X_test, y_train, y_test = train_test_split(X, y, test_size=0.2,
random_state=2017)
num_rounds = 100
```

```
kfold = model_selection.StratifiedKFold(n_splits=5, random_state=2017)

clf_XGB = XGBClassifier(n_estimators = num_rounds,
                        objective= 'binary:logistic',
                        seed=2017)

# use early_stopping_rounds to stop the cv when there is no score
imporovement
clf_XGB.fit(X_train,y_train, early_stopping_rounds=20, eval_set=[(X_test,
y_test)], verbose=False)

results = model_selection.cross_val_score(clf_XGB, X_train,y_train,
cv=kfold)
print ("\nxgBoost - CV Train : %.2f" % results.mean())
print ("xgBoost - Train : %.2f" % metrics.accuracy_score(clf_XGB.predict
(X_train), y_train))
print ("xgBoost - Test : %.2f" % metrics.accuracy_score(clf_XGB.predict
(X_test), y_test))
#----output----
xgBoost - CV Train : 0.69
xgBoost - Train : 0.73
xgBoost - Test : 0.74
```

Now let's also look at how to build a model using xgboost native interface. DMatrix the internal data structure of xgboostfor input data. It is good practice to convert the large dataset to DMatrix object to save preprocessing time (Listing 4-18).

Listing 4-18. xgboost Using It's Native Python Package Code

```
xgtrain = xgb.DMatrix(X_train, label=y_train, missing=-999)
xgtest = xgb.DMatrix(X_test, label=y_test, missing=-999)

# set xgboost params
param = {'max_depth': 3,  # the maximum depth of each tree
         'objective': 'binary:logistic'}
```

```
clf_xgb_cv = xgb.cv(param, xgtrain, num_rounds,
                    stratified=True,
                    nfold=5,
                    early_stopping_rounds=20,
                    seed=2017)

print ("Optimal number of trees/estimators is %i" % clf_xgb_cv.shape[0])

watchlist  = [(xgtest,'test'), (xgtrain,'train')]
clf_xgb = xgb.train(param, xgtrain,clf_xgb_cv.shape[0], watchlist)

# predict function will produce the probability
# so we'll use 0.5 cutoff to convert probability to class label
y_train_pred = (clf_xgb.predict(xgtrain, ntree_limit=clf_xgb.best_
iteration) > 0.5).astype(int)
y_test_pred = (clf_xgb.predict(xgtest, ntree_limit=clf_xgb.best_iteration)
> 0.5).astype(int)

print ("XGB - Train : %.2f" % metrics.accuracy_score(y_train_pred, y_train))
print ("XGB - Test : %.2f" % metrics.accuracy_score(y_test_pred, y_test))

    #----output----

Optimal number of trees (estimators) is 6
[0]     test-error:0.344156     train-error:0.299674
[1]     test-error:0.324675     train-error:0.273616
[2]     test-error:0.272727     train-error:0.281759
[3]     test-error:0.266234     train-error:0.278502
[4]     test-error:0.266234     train-error:0.273616
[5]     test-error:0.311688     train-error:0.254072
XGB - Train : 0.73
XGB - Test : 0.73
```

Ensemble Voting—Machine Learning's Biggest Heroes United

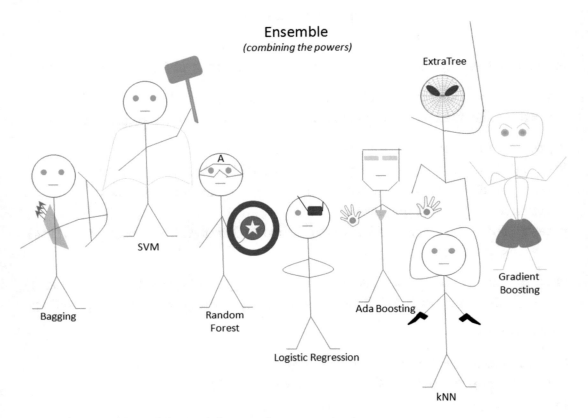

Figure 4-13. *Ensemble: ML's biggest heroes united*

A voting classifier enables us to combine the predictions through majority voting from multiple ML algorithms of different types, unlike Bagging/Boosting where a similar type of multiple classifiers is used for majority voting.

First, you can create multiple stand-alone models from your training dataset. Then, a voting classifier can be used to wrap your models and average the predictions of the submodels when asked to make predictions for new data. The predictions of the submodels can be weighted, but specifying the weights for classifiers manually or even heuristically is difficult. More advanced methods can learn how to best weight the predictions from submodels, but this is called stacking (stacked aggregation) and is currently not provided in Scikit-learn.

Let's build individual models on the Pima diabetes dataset and try the voting classifier, to combine model results to compare the change in accuracy (Listing 4-19).

Listing 4-19. Ensemble Model

```
import pandas as pd
import numpy as np

# set seed for reproducability
np.random.seed(2017)

import statsmodels.api as sm
from sklearn import metrics
from sklearn.linear_model import LogisticRegression
from sklearn.naive_bayes import GaussianNB
from sklearn.ensemble import RandomForestClassifier
from sklearn.svm import SVC
from sklearn.neighbors import KNeighborsClassifier
from sklearn.tree import DecisionTreeClassifier
from sklearn.ensemble import AdaBoostClassifier
from sklearn.ensemble import BaggingClassifier
from sklearn.ensemble import GradientBoostingClassifier

# currently its available as part of mlxtend and not sklearn
from mlxtend.classifier import EnsembleVoteClassifier
from sklearn import model_selection
from sklearn import metrics
from sklearn.model_selection import train_test_split

# read the data in
df = pd.read_csv("Data/Diabetes.csv")

X = df.iloc[:,:8]      # independent variables
y = df['class']        # dependent variables

# evaluate the model by splitting into train and test sets
X_train, X_test, y_train, y_test = train_test_split(X, y, test_size=0.3,
random_state=2017)
```

```
LR = LogisticRegression(random_state=2017)
RF = RandomForestClassifier(n_estimators = 100, random_state=2017)
SVM = SVC(random_state=0, probability=True)
KNC = KNeighborsClassifier()
DTC = DecisionTreeClassifier()
ABC = AdaBoostClassifier(n_estimators = 100)
BC = BaggingClassifier(n_estimators = 100)
GBC = GradientBoostingClassifier(n_estimators = 100)

clfs = []
print('5-fold cross validation:\n')
for clf, label in zip([LR, RF, SVM, KNC, DTC, ABC, BC, GBC],
                      ['Logistic Regression',
                       'Random Forest',
                       'Support Vector Machine',
                       'KNeighbors',
                       'Decision Tree',
                       'Ada Boost',
                       'Bagging',
                       'Gradient Boosting']):
    scores = model_selection.cross_val_score(clf, X_train, y_train, cv=5,
    scoring='accuracy')
    print("Train CV Accuracy: %0.2f (+/- %0.2f) [%s]" % (scores.mean(),
    scores.std(), label))
    md = clf.fit(X, y)
    clfs.append(md)
    print("Test Accuracy: %0.2f " % (metrics.accuracy_score(clf.predict
    (X_test), y_test)))
#----output----
5-fold cross validation:

Train CV Accuracy: 0.76 (+/- 0.03) [Logistic Regression]
Test Accuracy: 0.79
Train CV Accuracy: 0.74 (+/- 0.03) [Random Forest]
Test Accuracy: 1.00
Train CV Accuracy: 0.65 (+/- 0.00) [Support Vector Machine]
```

```
Test Accuracy: 1.00
Train CV Accuracy: 0.70 (+/- 0.05) [KNeighbors]
Test Accuracy: 0.84
Train CV Accuracy: 0.69 (+/- 0.02) [Decision Tree]
Test Accuracy: 1.00
Train CV Accuracy: 0.73 (+/- 0.04) [Ada Boost]
Test Accuracy: 0.83
Train CV Accuracy: 0.75 (+/- 0.04) [Bagging]
Test Accuracy: 1.00
Train CV Accuracy: 0.75 (+/- 0.03) [Gradient Boosting]
Test Accuracy: 0.92
```

From the previous benchmarking we see that 'Logistic Regression', 'Random Forest', 'Bagging', and Ada/Gradient Boosting algorithms give better accuracy compared with other models. Let's combine nonsimilar models such as logistic regression (base model), Random Forest (bagging model), and gradient boosting (boosting model) to create a robust generalized model.

Hard Voting vs. Soft Voting

Majority voting is also known as hard voting. The argmax of the sum of predicted probabilities is known as soft voting. Parameter "weights" can be used to assign specific weight to classifiers. The predicted class probabilities for each classifier are multiplied by the classifier weight and averaged. Then the final class label is derived from the highest average probability class label.

Assume we assign an equal weight of 1 to all classifiers (Table 4-1). Based on soft voting, the predicted class label is 1, as it has the highest average probability. Refer to Listing 4-20 for the example code implementation of the ensemble voting model.

Table 4-1. *Soft Voting*

Classifier	Class 1	Class 2	Class 3	.	.	Class n
Classifier 1	w1 * 0.3	w1 * 0.1	w1 * 0.6	.	.	w1 * 0.1
Classifier 2	w2 * 0.4	w2 * 0.3	w2 * 0.3	.	.	w2 * 0.3
Classifier 3	w3 * 0.5	w3 * 0.4	w3 * 0.2	.	.	w3 * 0.3
Weighted average	0.4	0.12	0.37	.	.	0.23

Note   Some classifiers of Scikit-learn do not support the predict_proba method.

Listing 4-20. Ensemble Voting Model

```
# ### Ensemble Voting
clfs = []
print('5-fold cross validation:\n')

ECH = EnsembleVoteClassifier(clfs=[LR, RF, GBC], voting='hard')
ECS = EnsembleVoteClassifier(clfs=[LR, RF, GBC], voting='soft',
weights=[1,1,1])

for clf, label in zip([ECH, ECS],
                      ['Ensemble Hard Voting',
                       'Ensemble Soft Voting']):
    scores = model_selection.cross_val_score(clf, X_train, y_train, cv=5,
    scoring='accuracy')
    print("Train CV Accuracy: %0.2f (+/- %0.2f) [%s]" % (scores.mean(),
    scores.std(), label))
    md = clf.fit(X, y)
    clfs.append(md)
    print("Test Accuracy: %0.2f " % (metrics.accuracy_score(clf.predict
    (X_test), y_test)))
#----output----
5-fold cross validation:

Train CV Accuracy: 0.75 (+/- 0.02) [Ensemble Hard Voting]
Test Accuracy: 0.93
Train CV Accuracy: 0.76 (+/- 0.02) [Ensemble Soft Voting]
Test Accuracy: 0.95
```

Stacking

David H. Wolpert presented (in 1992) the concept of stacked generalization, most commonly known as "stacking," in his publication with the *Neural Networks* journal. In stacking, initially you train multiple base models of a different type on a training/

test dataset. It is ideal to mix models that work differently (kNN, bagging, boosting, etc.) so they can learn some part of the problem. At level 1, use the predicted values from base models as features and train a model, which is known as the metamodel. Thus, combining the learning of individual model will result in improved accuracy. This is a simple level 1 stacking, and similarly, you can stack multiple levels of different types of models (Figure 4-14).

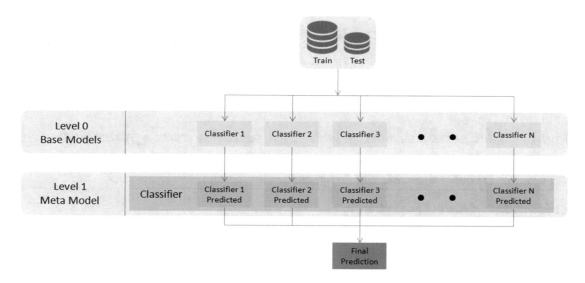

Figure 4-14. *Simple level 2 stacking model*

Let's apply the stacking concept previously discussed on the diabetes dataset and compare the accuracy of base vs. metamodel (Listing 4-21).

Listing 4-21. Model Stacking

```
# Classifiers
from sklearn.ensemble import RandomForestClassifier
from sklearn.ensemble import GradientBoostingClassifier
from sklearn.linear_model import LogisticRegression
from sklearn.neighbors import KNeighborsClassifier

import warnings
warnings.filterwarnings("ignore", category=FutureWarning)

seed = 2019
```

```
np.random.seed(seed)  # seed to shuffle the train set

# read the data in
df = pd.read_csv("Data/Diabetes.csv")

X = df.iloc[:,0:8] # independent variables
y = df['class'].values      # dependent variables

#Normalize
X = StandardScaler().fit_transform(X)

# evaluate the model by splitting into train and test sets
X_train, X_test, y_train, y_test = train_test_split(X, y, test_size=0.2,
random_state=seed)

kfold = model_selection.StratifiedKFold(n_splits=5, random_state=seed)
num_trees = 10
verbose = True # to print the progress

clfs = [KNeighborsClassifier(),
        RandomForestClassifier(n_estimators=num_trees, random_state=seed),
        GradientBoostingClassifier(n_estimators=num_trees, random_
        state=seed)]

# Creating train and test sets for blending
dataset_blend_train = np.zeros((X_train.shape[0], len(clfs)))
dataset_blend_test = np.zeros((X_test.shape[0], len(clfs)))

print('5-fold cross validation:\n')
for i, clf in enumerate(clfs):
    scores = model_selection.cross_val_score(clf, X_train, y_train,
    cv=kfold, scoring='accuracy')
    print("##### Base Model %0.0f #####" % i)
    print("Train CV Accuracy: %0.2f (+/- %0.2f)" % (scores.mean(),
    scores.std()))
    clf.fit(X_train, y_train)
    print("Train Accuracy: %0.2f " % (metrics.accuracy_score(clf.predict
    (X_train), y_train)))
```

```
        dataset_blend_train[:,i] = clf.predict_proba(X_train)[:, 1]
        dataset_blend_test[:,i] = clf.predict_proba(X_test)[:, 1]
        print("Test Accuracy: %0.2f " % (metrics.accuracy_score(clf.predict
        (X_test), y_test)))

print ("##### Meta Model #####")
clf = LogisticRegression()
scores = model_selection.cross_val_score(clf, dataset_blend_train, y_train,
cv=kfold, scoring='accuracy')
clf.fit(dataset_blend_train, y_train)
print("Train CV Accuracy: %0.2f (+/- %0.2f)" % (scores.mean(), scores.std()))
print("Train Accuracy: %0.2f " % (metrics.accuracy_score(clf.
predict(dataset_blend_train), y_train)))
print("Test Accuracy: %0.2f " % (metrics.accuracy_score(clf.
predict(dataset_blend_test), y_test)))
#----output----
5-fold cross validation:

##### Base Model 0 #####
Train CV Accuracy: 0.71 (+/- 0.03)
Train Accuracy: 0.83
Test Accuracy: 0.75
##### Base Model 1 #####
Train CV Accuracy: 0.73 (+/- 0.02)
Train Accuracy: 0.98
Test Accuracy: 0.79
##### Base Model 2 #####
Train CV Accuracy: 0.74 (+/- 0.01)
Train Accuracy: 0.80
Test Accuracy: 0.80
##### Meta Model #####
Train CV Accuracy: 0.99 (+/- 0.02)
Train Accuracy: 0.99
Test Accuracy: 0.77
```

Hyperparameter Tuning

One of the primary objectives and challenges in the ML process is improving the performance score, based on data patterns and observed evidence. To achieve this objective, almost all ML algorithms have a specific set of parameters that need to estimate from a dataset, which will maximize the performance score. Assume that these parameters are the knobs that you need to adjust to different values to find the optimal combination of parameters that give you the best model accuracy (Figure 4-15). The best way to choose a good hyperparameter is through trial and error of all possible combination of parameter values. Scikit-learn provides GridSearchCV and RandomSearchCV functions to facilitate an automatic and reproducible approach for hyperparameter tuning.

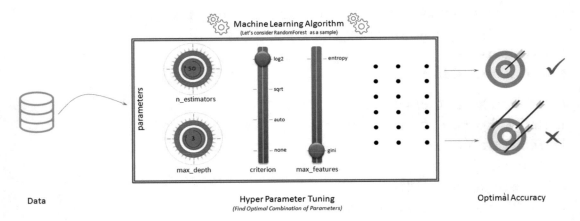

Figure 4-15. *Hyperparameter tuning*

GridSearch

For a given model, you can define a set of parameter values that you would like to try. Then, using the GridSearchCV function of Scikit-learn, models are built for all possible combinations of a preset list of values of hyperparameter provided by you and the best combination is chosen based on the cross-validation score. There are two disadvantages associated with GridSearchCV:

1. *Computationally expensive*: It is obvious that with more parameter values the GridSearch will be computationally expensive. Consider an example where you have five parameters, and

assume that you would like to try five values for each parameter, which will result in 5∗∗5 = 3,125 combinations. Further multiply this with a number of cross-validation folds being used (e.g., if k-fold is 5, then 3125 ∗ 5 = 15,625 model fits).

2. *Not perfectly optimal but nearly optimal parameters*: GridSearch will look at fixed points that you provide for the numerical parameters, hence there is a great chance of missing the optimal point that lies between the fixed points. For example, assume that you would like to try the fixed points for 'n_estimators': [100, 250, 500, 750, 1000] for a decision tree model and there is a chance that the optimal point might lie between the two fixed points. However, GridSearch is not designed to search between fixed points.

Let's try GridSearchCV for a RandomForest classifier on the Pima diabetes data set, to find the optimal parameter values (Listing 4-22).

Listing 4-22. Grid Search for Hyperparameter Tuning

```
from sklearn.ensemble import RandomForestClassifier
from sklearn.model_selection import GridSearchCV
seed = 2017

# read the data in
df = pd.read_csv("Data/Diabetes.csv")

X = df.iloc[:,:8].values      # independent variables
y = df['class'].values        # dependent variables

#Normalize
X = StandardScaler().fit_transform(X)

# evaluate the model by splitting into train and test sets
X_train, X_test, y_train, y_test = train_test_split(X, y, test_size=0.3,
random_state=seed)

kfold = model_selection.StratifiedKFold(n_splits=5, random_state=seed)
num_trees = 100

clf_rf = RandomForestClassifier(random_state=seed).fit(X_train, y_train)
```

```
rf_params = {
    'n_estimators': [100, 250, 500, 750, 1000],
    'criterion': ['gini', 'entropy'],
    'max_features': [None, 'auto', 'sqrt', 'log2'],
    'max_depth': [1, 3, 5, 7, 9]
}

# setting verbose = 10 will print the progress for every 10 task completion
grid = GridSearchCV(clf_rf, rf_params, scoring='roc_auc', cv=kfold,
verbose=10, n_jobs=-1)
grid.fit(X_train, y_train)

print ('Best Parameters: ', grid.best_params_)

results = model_selection.cross_val_score(grid.best_estimator_, X_train,y_
train, cv=kfold)
print ("Accuracy - Train CV: ", results.mean())
print ("Accuracy - Train : ", metrics.accuracy_score(grid.best_estimator_.
predict(X_train), y_train))
print ("Accuracy - Test : ", metrics.accuracy_score(grid.best_estimator_.
predict(X_test), y_test))
#----output----
Fitting 5 folds for each of 200 candidates, totalling 1000 fits
Best Parameters:  {'criterion': 'entropy', 'max_depth': 5, 'max_features':
'log2', 'n_estimators': 500}
Accuracy - Train CV:  0.7447905849775008
Accuracy - Train :  0.8621973929236499
Accuracy - Test :  0.7965367965367965
```

RandomSearch

As the name suggests, the RandomSearch algorithm tries random combinations of a range of values of given parameters. The numerical parameters can be specified as a range (unlike fixed values in GridSearch). You can control the number of iterations of random search that you would like to perform. It is known to find a very good combination in a lot less time compared with GridSearch; however, you have to carefully choose the range for parameters and the number of random search iterations, as it can miss the best parameter combination with lesser iteration or smaller range.

Let's try the RandomSearchCV for the same combination that we tried for GridSearch and compare the time/accuracy (Listing 4-23).

Listing 4-23. Random Search for Hyperparameter Tuning

```python
from sklearn.model_selection import RandomizedSearchCV
from scipy.stats import randint as sp_randint

# specify parameters and distributions to sample from
param_dist = {'n_estimators':sp_randint(100,1000),
              'criterion': ['gini', 'entropy'],
              'max_features': [None, 'auto', 'sqrt', 'log2'],
              'max_depth': [None, 1, 3, 5, 7, 9]
             }

# run randomized search
n_iter_search = 20
random_search = RandomizedSearchCV(clf_rf, param_distributions=param_dist,
                                   cv=kfold, n_iter=n_iter_search,
                                   verbose=10, n_jobs=-1, random_state=seed)

random_search.fit(X_train, y_train)
# report(random_search.cv_results_)

print ('Best Parameters: ', random_search.best_params_)

results = model_selection.cross_val_score(random_search.best_estimator_,
X_train,y_train, cv=kfold)
print ("Accuracy - Train CV: ", results.mean())
print ("Accuracy - Train : ", metrics.accuracy_score(random_search.best_
estimator_.predict(X_train), y_train))
print ("Accuracy - Test : ", metrics.accuracy_score(random_search.best_
estimator_.predict(X_test), y_test))
#----output----
Fitting 5 folds for each of 20 candidates, totalling 100 fits

Best Parameters:  {'criterion': 'entropy', 'max_depth': 3, 'max_features':
None, 'n_estimators': 694}
```

```
Accuracy - Train CV:  0.7542402215299411
Accuracy - Train :  0.7802607076350093
Accuracy - Test :  0.8051948051948052
```

Notice that in this case, with RandomSearchCV we were able to achieve comparable accuracy results with 100 fits to that of a GridSearchCV's 1000 fit.

Figure 4-16 is a sample illustration of how grid search vs. random search results differ (it's not the actual representation) between two parameters. Assume that the optimal area for max_depth lies between 3 and 5 (blue shade) and for n_estimators it lies between 500 and 700 (amber shade). The ideal optimal value for the combined parameter would lie where the individual regions intersect. Both methods will be able to find a nearly optimal parameter and not necessarily the perfect optimal point.

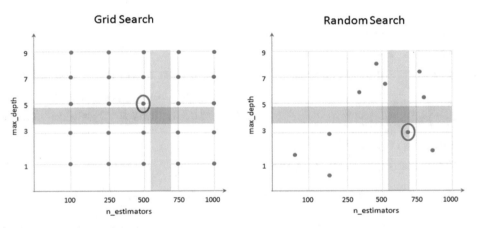

Figure 4-16. *Grid search vs. random search*

Bayesian Optimization

One of the key emerging hyperparameter tuning techniques is Bayesian optimization, using Gaussian process regression of the observed amalgamation of parameters and their associated target values. The objective of Bayesian optimization is to find the maximum value in as few iterations as possible for an unknown function. The key difference compared with grid and random search is that space has probability distributions for each hyperparameter rather than discrete values. This technique is particularly suited for optimization of high-cost functions, situations where the balance between exploration and exploitation is important. Although this technique works well for continuous variables, there is no intuitive way of dealing with discrete parameters. Refer to Listing 4-24 for a simple code implementation example for a random search for hyperparameter tuning.

You can learn more about the package and example at https://github.com/fmfn/BayesianOptimization

Listing 4-24. Random Search for Hyperparameter Tuning

```
# pip install bayesian-optimization
from bayes_opt import BayesianOptimization
from sklearn.model_selection import cross_val_score
from bayes_opt.util import Colours
from sklearn.ensemble import RandomForestClassifier as RFC

def rfc_cv(n_estimators, min_samples_split, max_features, data, targets):
    """Random Forest cross validation.
    This function will instantiate a random forest classifier with
    parameters
    n_estimators, min_samples_split, and max_features. Combined with
    data and
    targets this will in turn be used to perform cross-validation.
    The result of cross validation is returned. Our goal is to find
    combinations of n_estimators, min_samples_split, and
    max_features that minimzes the log loss.
    """
    estimator = RFC(
        n_estimators=n_estimators,
        min_samples_split=min_samples_split,
        max_features=max_features,
        random_state=2
    )
    cval = cross_val_score(estimator, data, targets,
                           scoring='neg_log_loss', cv=4)
    return cval.mean()
```

```python
def optimize_rfc(data, targets):
    """Apply Bayesian Optimization to Random Forest parameters."""
    def rfc_crossval(n_estimators, min_samples_split, max_features):
        """Wrapper of RandomForest cross validation.
        Notice how we ensure n_estimators and min_samples_split are casted
        to integer before we pass them along. Moreover, to avoid max_
        features
        taking values outside the (0, 1) range, we also ensure it is capped
        accordingly.
        """
        return rfc_cv(
            n_estimators=int(n_estimators),
            min_samples_split=int(min_samples_split),
            max_features=max(min(max_features, 0.999), 1e-3),
            data=data,
            targets=targets,
        )

    optimizer = BayesianOptimization(
        f=rfc_crossval,
        pbounds={
            "n_estimators": (10, 250),
            "min_samples_split": (2, 25),
            "max_features": (0.1, 0.999),
        },
        random_state=1234,
        verbose=2
    )
    optimizer.maximize(n_iter=10)

    print("Final result:", optimizer.max)
    return optimizer

print(Colours.green("--- Optimizing Random Forest ---"))
optimize_rfc(X_train, y_train)
#----output----
```

```
--- Optimizing Random Forest ---
|   iter    |  target   | max_fe... | min_sa... | n_esti... |
-------------------------------------------------------------
|    1      |  -0.5112  |  0.2722   |   16.31   |   115.1   |
|    2      |  -0.5248  |  0.806    |   19.94   |   75.42   |
|    3      |  -0.5075  |  0.3485   |   20.44   |   240.0   |
|    4      |  -0.528   |  0.8875   |   10.23   |   130.2   |
|    5      |  -0.5098  |  0.7144   |   18.39   |   98.86   |
|    6      |  -0.51    |  0.999    |   25.0    |   176.7   |
|    7      |  -0.5113  |  0.7731   |   24.94   |   249.8   |
|    8      |  -0.5339  |  0.999    |   2.0     |   250.0   |
|    9      |  -0.5107  |  0.9023   |   24.96   |   116.2   |
|   10      |  -0.8284  |  0.1065   |   2.695   |   10.04   |
|   11      |  -0.5235  |  0.1204   |   24.89   |   208.1   |
|   12      |  -0.5181  |  0.1906   |   2.004   |   81.15   |
|   13      |  -0.5203  |  0.1441   |   2.057   |   185.3   |
|   14      |  -0.5257  |  0.1265   |   24.85   |   153.1   |
|   15      |  -0.5336  |  0.9906   |   2.301   |   219.3   |
=============================================================
```

Final result: {'target': -0.5075247858575866, 'params': {'max_features': 0.34854136537364394, 'min_samples_split': 20.443060083305443, 'n_estimators': 239.95344488408924}}

Noise Reduction for Time-Series IoT Data

The last decade has seen a humongous growth in both the software and hardware side of technology, which gave birth to the concept of a connected world or Internet of Things (IoT). This means the physical devices, everyday objects and hardware forms, are fitted with tiny sensors to continuously capture the state of the machine in the form of different parameters, and extended to the Internet connectivity so that these devices are able to communicate/interact with each other and can be remotely monitored/controlled. Prognostic analytics is the ML area that deals with mining the IoT data to get insights. At a high level there are two key aspects. One is anomaly detection, a process to identify precursory failure signatures (unusual behavior also known as anomalies) as early as possible so that necessary action can be planned to avoid impeding failure—for example a sudden increase or decrease in voltage or temperature of a device. The other is remaining useful life (RUL) prediction. It is a process to predict RUL or how far away to the impending failure from the point of anomaly detection. We can use regression models to predict the RUL.

The data set that we collect from sensors is prone to noise, particularly the high bandwidth measurements such as vibration or ultrasound signals. The Fourier transform is one of the well-known techniques that allows us to perform analysis in the frequency or spectrum domain, to gain a deeper knowledge of our high bandwidth signals such as vibration measurement profile. Fourier is a series of sine waves and the Fourier transform is essentially deconstructing a signal into individual sine wave components. However, the drawback of the Fourier transform is that it cannot provide the local information if the spectral components of a signal change rapidly with time. The key drawback of the Fourier transform is that once a signal is transformed from the time domain to its frequency domain, all the information related to the time are lost. The wavelet transform addresses the key drawbacks of the Fourier transform, making it the ideal choice for high bandwidth signal processing. There is a wide array of literature that explains the wavelet transform and its application, so you won't get many details in this book. Essentially, wavelets can be used to decompose a signal into a series of coefficients. The first coefficients represent the lowest frequencies, and the last coefficients represent the highest frequencies. By removing the higher frequency coefficients and then reconstructing the signal with the truncated coefficients, we can smooth the signal without smoothing over all of the interesting peaks the way we would with a moving average.

Wavelet transforms decomposes time-series signal into two parts, a low frequency or low-pass filter that trends, smoothes the original signal approximations and a high frequency or high-pass filter that yields detailed local properties such as anomalies as shown in Figure 4-17.

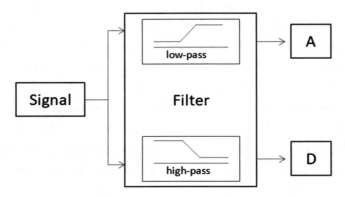

Figure 4-17. *Notion of filtering for wavelet decomposition*

One of the strengths of the wavelet transform is the ability to perform multi-level decomposition process by iterating to obtained multi-resolution, with the approximations being successively decomposed in turn, so that one signal can be broken down into many lower resolution components. The wavelet transform function f(x) is a sequence of numbers where wavelet basis ψ called a mother wavelet function is used for the decomposition.

$$f(x) = \frac{1}{\sqrt{M}} \sum_{k} W_{\phi}(j_0, k) \phi_{j_0, k}(x) + \frac{1}{\sqrt{M}} \sum_{j=j_0}^{\infty} \sum_{k} W_{\psi}(j, k) \psi_{j, k}(x)$$

Where, j_0 is an arbitrary starting scale called the approximation or scaling coefficient. $W_{\psi}(j,k)$ is called the detail or wavelet coefficients.

Listing 4-25. Wavelet Transform Implementation

```python
import pywt
from statsmodels.robust import mad
import pandas as pd
import numpy as np

df = pd.read_csv('Data/Temperature.csv')

# Function to denoise the sensor data using wavelet transform
def wp_denoise(df):
    for column in df:
        x = df[column]
        wp = pywt.WaveletPacket(data=x, wavelet='db7', mode='symmetric')
        new_wp = pywt.WaveletPacket(data=None, wavelet='db7', mode='sym')
        for i in range(wp.maxlevel):
            nodes = [node.path for node in wp.get_level(i, 'natural')]
            # Remove the high and low pass signals
            for node in nodes:
                sigma = mad(wp[node].data)
                uthresh = sigma * np.sqrt( 2*np.log( len( wp[node].data ) ) )
                new_wp[node] = pywt.threshold(wp[node].data, value=uthresh,
                mode='soft')
```

```
        y = new_wp.reconstruct(update=False)[:len(x)]
        df[column] = y
    return df

# denoise the sensor data
df_denoised = wp_denoise(df.iloc[:,3:4])
df['Date'] = pd.to_datetime(df['Date'])

plt.figure(1)
ax1 = plt.subplot(221)
df['4030CFDC'].plot(ax=ax1, figsize=(8, 8), title='Signal with noise')

ax2 = plt.subplot(222)
df_denoised['4030CFDC'].plot(ax=ax2, figsize=(8, 8), title='Signal
without noise')
plt.tight_layout()
#----output----
```

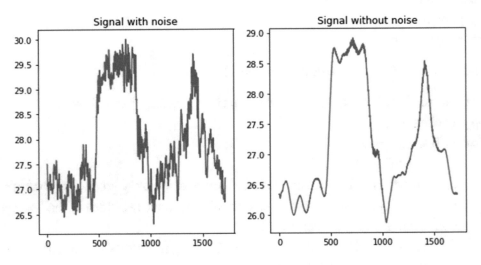

Summary

In this step, we have learned various common issues that can hinder model accuracy, such as not choosing the optimal probability cutoff point for class creation, variance, and bias. We also briefly looked at different model tuning techniques practiced, such as bagging, boosting, ensemble voting, and grid search/random search and Bayesian optimization techniques for hyperparameter tuning. We also looked at noise reduction technique for IoT data. To be concise, we only looked at the most important aspects for each of the topics discussed to get you started. However, there are more options for each of the algorithms for tuning, and each of these techniques has been evolving at a fast pace. So I encourage you to keep an eye on their respective officially hosted webpages and GitHub repository (Table 4-2).

Table 4-2. *Additional Resources*

Name	Web Page	Github Repository
Scikit-learn	http://scikit-learn.org/stable/#	https://github.com/scikit-learn/scikit-learn
Xgboost	https://xgboost.readthedocs.io/en/latest/	https://github.com/dmlc/xgboost
Bayesian optimization	N/A	https://github.com/fmfn/BayesianOptimization
Wavelet transforms	https://pywavelets.readthedocs.io/en/latest/#	https://github.com/PyWavelets/pywt

We have reached the end of step 4, which means you have already passed halfway through your machine learning journey. In the next chapter, we'll learn text mining techniques.

CHAPTER 5

Step 5: Text Mining and Recommender Systems

One of the key areas of artificial intelligence is natural language processing (NLP), or text mining as it's generally known, which deals with teaching computers how to extract meaning from text. Over the last 2 decades, with the explosion of the Internet world and the rise of social media, there is plenty of valuable data being generated in the form of text. The process of unearthing meaningful patterns from text data is called text mining. This chapter covers an overview of the high-level text mining process, key concepts, and common techniques involved.

Apart from Scikit-learn, there are many established NLP-focused libraries available for Python, and the number has been growing over time. Table 5-1 lists the most popular libraries, based on their number of contributors as of 2016.

Table 5-1. *Popular Python Text Mining Libraries*

Package Name	# of Contributors (2019)	License	Description
NLTK	255	Apache	It's most popular and widely used toolkit predominantly built to support research and development of NLP
Gensim	311	LGPL-2	Mainly built for large corpus topic modeling, document indexing, and similarity retrieval
spaCy	300	MIT	Built using Python + Cython for efficient production implementation of NLP concepts

(continued)

© Manohar Swamynathan 2019
M. Swamynathan, *Mastering Machine Learning with Python in Six Steps*,
https://doi.org/10.1007/978-1-4842-4947-5_5

Table 5-1. (*continued*)

Package Name	# of Contributors (2019)	License	Description
Textblob	36	MIT	It's a wrapper around NLTK and Pattern libraries for easy accessibility of their capabilities. Suitable for fast prototyping
Polyglot	22	GPL-3	This is a multilingual text processing toolkit and supports massive multilingual applications.
Pattern	19	BSD-3	It's a web mining module for Python with capabilities included for scraping, NLP, machine learning, and network analysis/visualization.

Note   Another well-known library is Stanford CoreNLP, a suite of Java-based toolkits. There are a number of Python wrappers available for the same; however, the number of contributors for these wrappers is on the low side as of now.

Text Mining Process Overview

The overall text mining process can be broadly categorized into the following four phases, as shown in Figure 5-1:

1. Text data assemble

2. Text data preprocessing

3. Data exploration or visualization

4. Model building

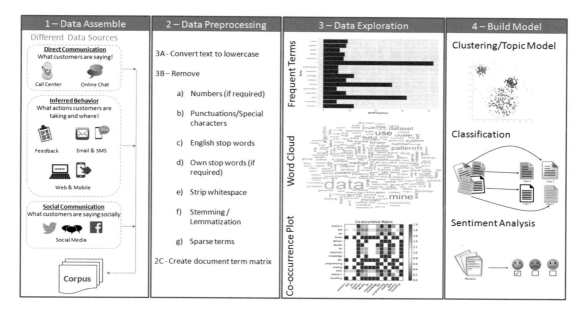

Figure 5-1. *Text mining process overview*

Data Assemble (Text)

It is observed that 70% of data available to any businesses is unstructured. The first step is collating unstructured data from different sources such as open-ended feedback; phone calls; e-mail support; online chat; and social media networks like Twitter, LinkedIn, and Facebook. Assembling these data and applying mining/machine learning (ML) techniques to analyze them provides valuable opportunities for organizations to build more power into the customer experience.

There are several libraries available for extracting text content from the different formats discussed. By far the best library that provides a simple, single interface for multiple formats is "textract" (open source MIT license). Note that as of now this library/package is available for Linux and Mac OS but not Windows. Table 5-2 lists the supported formats.

Table 5-2. *textract-Supported Formats*

Format	Supported Via	Additional Info
.csv / .eml / .json / .odt / .txt /	Python built-ins	
.doc	Antiword	www.winfield.demon.nl/
.docx	Python-docx	https://python-docx.readthedocs.io/en/latest/
.epub	Ebooklib	https://github.com/aerkalov/ebooklib
.gif / .jpg / .jpeg / .png / .tiff / .tif	tesseract-ocr	https://github.com/tesseract-ocr
.html / .htm	Beautifulsoup4	http://beautiful-soup-4.readthedocs.io/en/latest/
.mp3 / .ogg / .wav	SpeechRecongnition and sox	URL 1: https://pypi.python.org/pypi/SpeechRecognition/ URL 2: http://sox.sourceforge.net/
.msg	msg-extractor	https://github.com/mattgwwalker/msg-extractor
.pdf	pdftotext and pdfminer.six	URL 1: https://poppler.freedesktop.org/ URL 2: https://github.com/pdfminer/pdfminer.six
.pptx	Python-pptx	https://python-pptx.readthedocs.io/en/latest/
.ps	ps2text	http://pages.cs.wisc.edu/~ghost/doc/pstotext.htm
.rtf	Unrtf	www.gnu.org/software/unrtf/
.xlsx / .xls	Xlrd	https://pypi.python.org/pypi/xlrd

Let's look at the code for the most widespread formats in the business world: pdf, jpg, and audio files (Listing 5-1). Note that extracting text from other formats is also relatively simple.

Listing 5-1. Example Code for Extracting Data from pdf, jpg, Audio

```
# You can read/learn more about latest updates about textract on their
official documents site at http://textract.readthedocs.io/en/latest/
import textract

# Extracting text from normal pdf
text = textract.process('Data/PDF/raw_text.pdf', language='eng')

# Extrcting text from two columned pdf
text = textract.process('Data/PDF/two_column.pdf', language='eng')

# Extracting text from scanned text pdf
text = textract.process('Data/PDF/ocr_text.pdf', method='tesseract',
language='eng')

# Extracting text from jpg
text = textract.process('Data/jpg/raw_text.jpg', method='tesseract',
language='eng')

# Extracting text from audio file
text = textract.process('Data/wav/raw_text.wav', language='eng')
```

Social Media

Did you know that Twitter, the online news and social networking service provider, has 320 million users, with an average of 42 million active Tweets every day! *(Source: Global social media research summary 2016 by smart insights)*

Let's understand how to explore the rich information of social media (I'll consider Twitter as an example) to explore what is being spoken about a chosen topic (Figure 5-2). Most of these forums provide API for developers to access the posts.

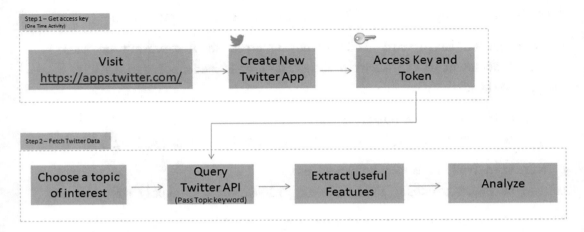

Figure 5-2. *Pulling Twitter posts for analysis*

Step 1—Get an access key (one-time activity). Take the following steps to set up a new Twitter app to get a consumer/access key, secret, and token (do not share the key token with unauthorized persons).

1. Go to `https://apps.twitter.com/`

2. Click "Create New App"

3. Fill the required information and click "Create your Twitter Application"

4. You'll get the access details under the "Keys and Access Tokens" tab

Step 2—Fetching tweets. Once you have the authorization secret and access tokens, you can use the Listing 5-2 code example to establish the connection.

Listing 5-2. Twitter Authentication

```
#Import the necessary methods from tweepy library
import tweepy
from tweepy.streaming import StreamListener
from tweepy import OAuthHandler
from tweepy import Stream
```

```
#provide your access details below
access_token = "Your token goes here"
access_token_secret = "Your token secret goes here"
consumer_key = "Your consumer key goes here"
consumer_secret = "Your consumer secret goes here"

# establish a connection
auth = tweepy.auth.OAuthHandler(consumer_key, consumer_secret)
auth.set_access_token(access_token, access_token_secret)

api = tweepy.API(auth)
```

Let's assume that you would like to understand what is being talked about concerning the iPhone 7 and its camera feature. So, let's pull ten recent posts.

Note: You can pull historic user posts about a topic for a max of 10 to 15 days only, depending on the volume of the posts.

```
#fetch recent 10 tweets containing words iphone7, camera
fetched_tweets = api.search(q=['iPhone 7','iPhone7','camera'], result_
type='recent', lang='en', count=10)
print ("Number of tweets: ",len(fetched_tweets))
#----output----
Number of tweets:  10

# Print the tweet text
for tweet in fetched_tweets:
    print ('Tweet ID: ', tweet.id)
    print ('Tweet Text: ', tweet.text, '\n')
#----output----
Tweet ID:  825155021390049281
Tweet Text:  RT @volcanojulie: A Tau Emerald dragonfly. The iPhone 7 camera
is exceptional!
#nature #insect #dragonfly #melbourne #australia #iphone7 #…

Tweet ID:  825086303318507520
Tweet Text:  Fuzzy photos? Protect your camera lens instantly with #iPhone7
Full Metallic Case. Buy now! https://t.co/d0dX4OBHL6 https://t.co/AInlBoreht
```

You can capture useful features onto a dataframe for further analysis (Listing 5-3).

Listing 5-3. Save Features to Dataframe

```
# function to save required basic tweets info to a dataframe
def populate_tweet_df(tweets):
    #Create an empty dataframe
    df = pd.DataFrame()

    df['id'] = list(map(lambda tweet: tweet.id, tweets))
    df['text'] = list(map(lambda tweet: tweet.text, tweets))
    df['retweeted'] = list(map(lambda tweet: tweet.retweeted, tweets))
    df['place'] = list(map(lambda tweet: tweet.user.location, tweets))
    df['screen_name'] = list(map(lambda tweet: tweet.user.screen_name, tweets))
    df['verified_user'] = list(map(lambda tweet: tweet.user.verified, tweets))
    df['followers_count'] = list(map(lambda tweet: tweet.user.followers_
    count, tweets))
    df['friends_count'] = list(map(lambda tweet: tweet.user.friends_count,
    tweets))

    # Highly popular user's tweet could possibly seen by large audience, so
    lets check the popularity of user
    df['friendship_coeff'] = list(map(lambda tweet: float(tweet.user.
    followers_count)/float(tweet.user.friends_count), tweets))
    return df

df = populate_tweet_df(fetched_tweets)
print df.head(10)
#---output----
                      id                                              text
0   825155021390049281   RT @volcanojulie: A Tau Emerald dragonfly. The...
1   825086303318507520   Fuzzy photos? Protect your camera lens instant...
2   825064476714098690   RT @volcanojulie: A Tau Emerald dragonfly. The...
3   825062644986023936   RT @volcanojulie: A Tau Emerald dragonfly. The...
4   824935025217040385   RT @volcanojulie: A Tau Emerald dragonfly. The...
5   824933631365779458   A Tau Emerald dragonfly. The iPhone 7 camera i...
6   824836880491483136   The camera on the IPhone 7 plus is fucking awe...
```

```
7  823805101999390720  'Romeo and Juliet' Ad Showcases Apple's iPhone...
8  823804251117850624  iPhone 7 Images Show Bigger Camera Lens - I ha...
9  823689306376196096  RT @computerworks5: Premium HD Selfie Stick &a...
```

	retweeted	place	screen_name	verified_user
0	False	Melbourne, Victoria	MonashEAE	False
1	False	California, USA	ShopNCURV	False
2	False	West Islip, Long Island, NY	FusionWestIslip	False
3	False	6676 Fresh Pond Rd Queens, NY	FusionRidgewood	False
4	False		Iphone7review	False
5	False	Melbourne; Monash University	volcanojulie	False
6	False	Hollywood, FL	Hbk_Mannyp	False
7	False	Toronto.NYC.the Universe	AaronRFernandes	False
8	False	Lagos, Nigeria	moyinoluwa_mm	False
9	False		Iphone7review	False

	followers_count	friends_count	friendship_coeff
0	322	388	0.829897
1	279	318	0.877358
2	13	193	0.067358
3	45	218	0.206422
4	199	1787	0.111360
5	398	551	0.722323
6	57	64	0.890625
7	18291	7	2613.000000
8	780	302	2.582781
9	199	1787	0.111360

Instead of a topic, you can also choose a screen_name focused on a topic. Let's look (Listing 5-4) at the posts by the screen name Iphone7review.

Listing 5-4. Example Code for Extracting Tweets Based on Screen Name

```
# For help about api look here http://tweepy.readthedocs.org/en/v2.3.0/api.html
fetched_tweets =  api.user_timeline(id='Iphone7review', count=5)
```

```
# Print the tweet text
for tweet in fetched_tweets:
    print 'Tweet ID: ', tweet.id
    print 'Tweet Text: ', tweet.text, '\n'
#----output----
Tweet ID:  825169063676608512
Tweet Text:  RT @alicesttu: iPhone 7S to get Samsung OLED display next year
#iPhone https://t.co/BylKbvXgAG #iphone

Tweet ID:  825169047138533376
Tweet Text:  Nothing beats the Iphone7! Who agrees? #Iphone7 https://t.co/
e03tXeLOao
```

Glancing through the posts quickly, one can generally conclude that there are positive comments about the camera features of the iPhone 7.

Data Preprocessing (Text)

This step deals with cleansing the consolidated text to remove noise, to ensure efficient syntactic, semantic text analysis for deriving meaningful insights from the text. Some common cleaning steps are briefed described in the following.

Convert to Lower Case and Tokenize

Here, all the data is converted into lower case. This is carried out to prevent words like "LIKE" or "Like" being interpreted as different words. Python provides a function *lower()* to convert text to lowercase.

Tokenizing is the process of breaking a large set of text into smaller meaningful chunks such as sentences, words, phrases.

Sentence Tokenizing

The NLTK (Natural Language Toolkit) library provides sent_tokenize for sentence level tokenizing, which uses a pretrained model, PunktSentenceTokenize, to determine punctuation and characters marking the end of the sentence for European languages (Listing 5-5).

Listing 5-5. Example Code for Sentence Tokenizing

```
import nltk
from nltk.tokenize import sent_tokenize

text='Statistics skills, and programming skills are equally important
for analytics. Statistics skills and domain knowledge are important for
analytics. I like reading books and traveling.'

sent_tokenize_list = sent_tokenize(text)
print(sent_tokenize_list)
#----output----
['Statistics skills, and programming skills are equally important for
analytics.', 'Statistics skills, and domain knowledge are important for
analytics.', 'I like reading books and travelling.']
```

There is a total of 17 European languages that NLTK supports for sentence tokenize. Listing 5-6 gives you the example code for you to load the tokenized model for specific language, saved as a pickle file as part of nltk.data

Listing 5-6. Sentence Tokenizing for European Languages

```
import nltk.data
spanish_tokenizer = nltk.data.load('tokenizers/punkt/spanish.pickle')
spanish_tokenizer.tokenize('Hola. Esta es una frase espanola.')
#----output----
['Hola.', 'Esta es una frase espanola.']
```

Word Tokenizing

The word_tokenize function of NLTK is a wrapper function that calls tokenize by the TreebankWordTokenizer (Listing 5-7).

Listing 5-7. Example Code for Word Tokenizing

```
from nltk.tokenize import word_tokenize
print word_tokenize(text)
```

335

```
# Another equivalent call method using TreebankWordTokenizer
from nltk.tokenize import TreebankWordTokenizer
tokenizer = TreebankWordTokenizer()
print (tokenizer.tokenize(text))
#----output----
['Statistics', 'skills', ',', 'and', 'programming', 'skills', 'are',
'equally', 'important', 'for', 'analytics', '.', 'Statistics', 'skills',
',', 'and', 'domain', 'knowledge', 'are', 'important', 'for', 'analytics',
'.', 'I', 'like', 'reading', 'books', 'and', 'travelling', '.']
```

Removing Noise

You should remove all information that is not comparative or relevant to text analytics. This can be seen as noise to the text analytics. Most common noises are numbers, punctuations, stop words, white space, etc. (Listing 5-8).

Numbers: Numbers are removed, as they may not be relevant and not hold valuable information.

Listing 5-8. Example Code for Removing Noise from Text

```
def remove_numbers(text):
    return re.sub(r'\d+', '', text)

text = 'This is a    sample English    sentence, \n with whitespace and
numbers 1234!'
print ('Removed numbers: ', remove_numbers(text))
#----output----
Removed numbers:  This is a    sample English    sentence,
 with whitespace and numbers!
```

Punctuation: It is to be removed to better identify each word and remove punctuation characters from the data set. For example: "like," and "like" or "coca-cola" and "CocaCola" would be interpreted as different words if the punctuation was not removed (Listing 5-9).

Listing 5-9. Example Code for Removing Punctuations from Text

```
import string
# Function to remove punctuations
def remove_punctuations(text):
    words = nltk.word_tokenize(text)
    punt_removed = [w for w in words if w.lower() not in string.
    punctuation]
    return " ".join(punt_removed)

print (remove_punctuations('This is a sample English sentence, with
punctuations!'))
#----output----
This is a sample English sentence with punctuations
```

Stop words: Words like "the," "and," and "or" are uninformative and add unneeded noise to the analysis. For this reason, they are removed (Listing 5-10).

Listing 5-10. Example Code for Removing Stop Words from the Text

```
from nltk.corpus import stopwords

# Function to remove stop words
def remove_stopwords(text, lang='english'):
    words = nltk.word_tokenize(text)
    lang_stopwords = stopwords.words(lang)
    stopwords_removed = [w for w in words if w.lower() not in lang_
    stopwords]
    return " ".join(stopwords_removed)

print (remove_stopwords('This is a sample English sentence'))
#----output----
sample English sentence
```

Note   *Remove own stop words (if required).* Certain words could be very commonly used in a particular domain. Along with English stop words, we could instead or in addition remove our own stop words. The choice of our own stop words might depend on the domain of discourse and might not become apparent until we've done some analysis.

Whitespace: Often in text analytics, extra whitespace (space, tab, carriage return, line feed) becomes identified as a word. This anomaly is avoided through a basic programming procedure in this step (Listing 5-11).

Listing 5-11. Example Code for Removing Whitespace from Text

```
# Function to remove whitespace
def remove_whitespace(text):
    return " ".join(text.split())
text = 'This is a     sample  English   sentence, \n with whitespace and
numbers 1234!'
print ('Removed whitespace: ', remove_whitespace(text))
#----output----
Removed whitespace:  This is a sample English sentence, with whitespace and
numbers 1234!
```

Part of Speech (PoS) Tagging

PoS tagging is the process of assigning language-specific parts of speech such as nouns, verbs, adjectives, adverbs, etc., for each word in the given text.

NLTK supports multiple PoS tagging models, and the default tagger is maxent_treebank_pos_tagger, which uses the Penn (Pennsylvania University) Treebank corpus (Table 5-3). The same has 36 possible PoS tags. A sentence (S) is represented by the parser as a tree having three children: a noun phrase (NP), a verbal phrase (VP), and the full stop (.). The root of the tree will be S. Listings 5-12 and 5-13 provide you example code for PoS tagging and visualizing the sentence tree.

Table 5-3. *NLTK PoS Taggers*

PoS Tagger	Short Description
maxent_treebank_pos_tagger	It's based on Maximum Entropy (ME) classification principles trained on Wall Street Journal subset of the Penn Treebank corpus
BrillTagger	Brill's transformational rule-based tagger
CRFTagger	Conditional random fields
HiddenMarkovModelTagger	Hidden Markov Models (HMMs) largely used to assign the correct label sequence to sequential data or assess the probability of a given label and data sequence
HunposTagge	A module for interfacing with the HunPos open-source PoS tagger
PerceptronTagger	Based on the averaged perceptron technique proposed by Matthew Honnibal
SennaTagger	Semantic/syntactic extraction using a neural network architecture
SequentialBackoffTagger	Classes for tagging sentences sequentially left to right
StanfordPOSTagger	Researched and developed at Stanford University
TnT	Implementation of "TnT — A Statistical Part of Speech Tagger" by Thorsten Brants

Listing 5-12. Example Code for PoS Tagging the Sentence and Visualizing the Sentence Tree

```
from nltk import chunk

tagged_sent = nltk.pos_tag(nltk.word_tokenize('This is a sample English
sentence'))
print (tagged_sent)

tree = chunk.ne_chunk(tagged_sent)
tree.draw() # this will draw the sentence tree
#----output----
[('This', 'DT'), ('is', 'VBZ'), ('a', 'DT'), ('sample', 'JJ'),
('English', 'JJ'), ('sentence', 'NN')]
```

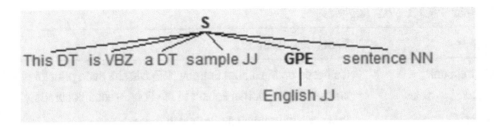

Listing 5-13. Example Code for Using Perceptron Tagger and Getting Help on Tags

```
# To use PerceptronTagger
from nltk.tag.perceptron import PerceptronTagger
PT = PerceptronTagger()
print (PT.tag('This is a sample English sentence'.split()))
#----output----
[('This', 'DT'), ('is', 'VBZ'), ('a', 'DT'), ('sample', 'JJ'), ('English', 'JJ'),
('sentence', 'NN')]

# To get help about tags
nltk.help.upenn_tagset('NNP')
#----output----
NNP: noun, proper, singular
```

Stemming

Stemming is the process of transforming to the root word. It uses an algorithm that removes common word endings for English words, such as "ly," "es," "ed," and "s." For example, assuming for an analysis you may want to consider "carefully," "cared", "cares," and "caringly" as "care" instead of separate words. The three most widely used stemming algorithms are listed in Figure 5-3. Listing 5-14 provides an example code for stemming.

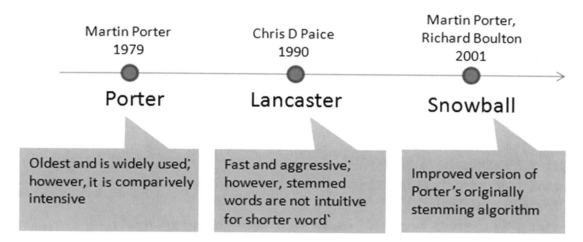

Figure 5-3. *Most popular NLTK stemmers*

Listing 5-14. Example Code for Stemming

```
from nltk import PorterStemmer, LancasterStemmer, SnowballStemmer

# Function to apply stemming to a list of words
def words_stemmer(words, type="PorterStemmer", lang="english",
encoding="utf8"):
    supported_stemmers = ["PorterStemmer","LancasterStemmer",
    "SnowballStemmer"]
    if type is False or type not in supported_stemmers:
        return words
    else:
        stem_words = []
        if type == "PorterStemmer":
            stemmer = PorterStemmer()
            for word in words:
                stem_words.append(stemmer.stem(word).encode(encoding))
        if type == "LancasterStemmer":
            stemmer = LancasterStemmer()
            for word in words:
                stem_words.append(stemmer.stem(word).encode(encoding))
```

```
        if type == "SnowballStemmer":
            stemmer = SnowballStemmer(lang)
            for word in words:
                stem_words.append(stemmer.stem(word).encode(encoding))
        return " ".join(stem_words)

words =  'caring cares cared caringly carefully'

print ("Original: ", words)
print ("Porter: ", words_stemmer(nltk.word_tokenize(words),
"PorterStemmer"))
print ("Lancaster: ", words_stemmer(nltk.word_tokenize(words),
"LancasterStemmer"))
print ("Snowball: ", words_stemmer(nltk.word_tokenize(words),
"SnowballStemmer"))
#----output----
Original:  caring cares cared caringly carefully
Porter:  care care care caringly care
Lancaster:  car car car car car
Snowball:  care care care care care
```

Lemmatization

It is the process of transforming to the dictionary base form. For this you can use WordNet, which is a large lexical database for English words that are linked together by their semantic relationships. It works as a thesaurus: it groups words together based on their meanings (Listing 5-15).

Listing 5-15. Example Code for Lemmatization

```
from nltk.stem import WordNetLemmatizer

wordnet_lemmatizer = WordNetLemmatizer()

# Function to apply lemmatization to a list of words
def words_lemmatizer(text, encoding="utf8"):
    words = nltk.word_tokenize(text)
    lemma_words = []
```

```
    wl = WordNetLemmatizer()
    for word in words:
        pos = find_pos(word)
        lemma_words.append(wl.lemmatize(word, pos).encode(encoding))
    return " ".join(lemma_words)

# Function to find part of speech tag for a word
def find_pos(word):
    # Part of Speech constants
    # ADJ, ADJ_SAT, ADV, NOUN, VERB = 'a', 's', 'r', 'n', 'v'
    # You can learn more about these at http://wordnet.princeton.edu/
    wordnet/man/wndb.5WN.html#sect3
    # You can learn more about all the penn tree tags at https://www.ling.
    upenn.edu/courses/Fall_2003/ling001/penn_treebank_pos.html
    pos = nltk.pos_tag(nltk.word_tokenize(word))[0][1]
    # Adjective tags - 'JJ', 'JJR', 'JJS'
    if pos.lower()[0] == 'j':
        return 'a'
    # Adverb tags - 'RB', 'RBR', 'RBS'
    elif pos.lower()[0] == 'r':
        return 'r'
    # Verb tags - 'VB', 'VBD', 'VBG', 'VBN', 'VBP', 'VBZ'
    elif pos.lower()[0] == 'v':
        return 'v'
    # Noun tags - 'NN', 'NNS', 'NNP', 'NNPS'
    else:
        return 'n'

print ("Lemmatized: ", words_lemmatizer(words))
#----output----
Lemmatized:  care care care caringly carefully
```

In the preceding case,'caringly'/'carefully' are inflected forms of care and they are an entry word listed in WordNet Dictionary so they are retained in their actual form itself.

NLTK English WordNet includes approximately 155,287 words and 117,000 synonym sets. For a given word, WordNet includes/provides a definition, example, synonyms (a group of nouns, adjectives, verbs that are similar), antonyms (opposite in meaning to another) etc. Listing 5-16 provies an example code for wordnet.

Listing 5-16. Example Code for Wordnet

```
from nltk.corpus import wordnet

syns = wordnet.synsets("good")
print "Definition: ", syns[0].definition()
print "Example: ", syns[0].examples()

synonyms = []
antonyms = []

# Print  synonums and antonyms (having opposite meaning words)
for syn in wordnet.synsets("good"):
    for l in syn.lemmas():
        synonyms.append(l.name())
        if l.antonyms():
            antonyms.append(l.antonyms()[0].name())

print ("synonyms: \n", set(synonyms))
print ("antonyms: \n", set(antonyms))
#----output----
Definition:  benefit
Example:  [u'for your own good', u"what's the good of worrying?"]
synonyms:
set([u'beneficial', u'right', u'secure', u'just', u'unspoilt',
u'respectable', u'good', u'goodness', u'dear', u'salutary', u'ripe',
u'expert', u'skillful', u'in_force', u'proficient', u'unspoiled',
u'dependable', u'soundly', u'honorable', u'full', u'undecomposed', u'safe',
u'adept', u'upright', u'trade_good', u'sound', u'in_effect', u'practiced',
u'effective', u'commodity', u'estimable', u'well', u'honest', u'near',
u'skilful', u'thoroughly', u'serious'])
antonyms:
set([u'bad', u'badness', u'ill', u'evil', u'evilness'])
```

N-grams

One of the important concepts in text mining is n-grams, which are fundamentally a set of cooccurring or continuous sequence of n items from a given sequence of large text. The items here could be words, letters, and syllables. Let's consider a sample sentence and try to extract n-grams for different values of n (Listing 5-17).

Listing 5-17. Example Code for Extracting n-grams from the Sentence

```
from nltk.util import ngrams
from collections import Counter

# Function to extract n-grams from text
def get_ngrams(text, n):
    n_grams = ngrams(nltk.word_tokenize(text), n)
    return [ ' '.join(grams) for grams in n_grams]

text = 'This is a sample English sentence'
print ("1-gram: ", get_ngrams(text, 1))
print ("2-gram: ", get_ngrams(text, 2))
print ("3-gram: ", get_ngrams(text, 3))
print ("4-gram: ", get_ngrams(text, 4))
#----output----
1-gram:['This', 'is', 'a', 'sample', 'English', 'sentence']
2-gram:['This is', 'is a', 'a sample', 'sample English', 'English sentence']
3-gram:['This is a', 'is a sample', 'a sample English', 'sample English sentence']
4-gram: ['This is a sample', 'is a sample English', 'a sample English sentence']
```

Note   1-gram is also called a unigram; 2-gram and 3-gram are bigram and trigram, respectively.

N-gram technique is relatively simple, and simply increasing the value of n will give us more contexts. It is widely used in probabilistic language models for predicting the next item in a sequence. For example, search engines use this technique to predict/ recommend the possibility of next character/words in the sequence for the user as they type (Listing 5-18).

Listing 5-18. Example Code for Extracting 2-grams from the Sentence and Storing in a Dataframe

```
text = 'Statistics skills, and programming skills are equally important for
analytics. Statistics skills and domain knowledge are important for analytics'

# remove punctuations
text = remove_punctuations(text)

# Extracting bigrams
result = get_ngrams(text,2)

# Counting bigrams
result_count = Counter(result)

# Converting the result to a data frame
import pandas as pd
df = pd.DataFrame.from_dict(result_count, orient='index')
df = df.rename(columns={'index':'words', 0:'frequency'}) # Renaming index
and column name
print (df)
#----output----
                      frequency
are equally               1
domain knowledge          1
skills are                1
knowledge are             1
programming skills        1
are important             1
skills and                2
for analytics             2
and domain                1
important for             2
and programming           1
Statistics skills         2
equally important         1
analytics Statistics      1
```

Bag of Words

The texts have to be represented as numbers to be able to apply any algorithms. Bag of words (BoW) is the method where you count the occurrence of words in a document without giving importance to the grammar and the order of words. This can be achieved by creating the Term-Document Matrix (TDM). It is simply a matrix with terms as the rows, document names as the columns, and a count of the frequency of words as the cells of the matrix (Figure 5-4). Let's learn to create TDM through an example: consider three text documents with some text in them. Sklearn provides good function under feature_extraction.text to convert a collection of text documents to the matrix of word counts (Listing 5-19).

Listing 5-19. Creating a Term Document Matrix from a Corpus of Sample Documents

```
import os
import pandas as pd
from sklearn.feature_extraction.text import CountVectorizer

# Function to create a dictionary with key as file names and values as text
for all files in a given folder
def CorpusFromDir(dir_path):
    result = dict(docs = [open(os.path.join(dir_path,f)).read() for f in
    os.listdir(dir_path)],
                  ColNames = map(lambda x: x, os.listdir(dir_path)))
    return result

docs = CorpusFromDir('Data/')

# Initialize
vectorizer = CountVectorizer()
doc_vec = vectorizer.fit_transform(docs.get('docs'))

#create dataFrame
df = pd.DataFrame(doc_vec.toarray().transpose(), index = vectorizer.get_
feature_names())
```

```
# Change column headers to be file names
df.columns = docs.get('ColNames')
print (df)
#----output----
```

	Doc_1.txt	Doc_2.txt	Doc_3.txt
analytics	1	1	0
and	1	1	1
are	1	1	0
books	0	0	1
domain	0	1	0
equally	1	0	0
for	1	1	0
important	1	1	0
knowledge	0	1	0
like	0	0	1
programming	1	0	0
reading	0	0	1
skills	2	1	0
statistics	1	1	0
travelling	0	0	1

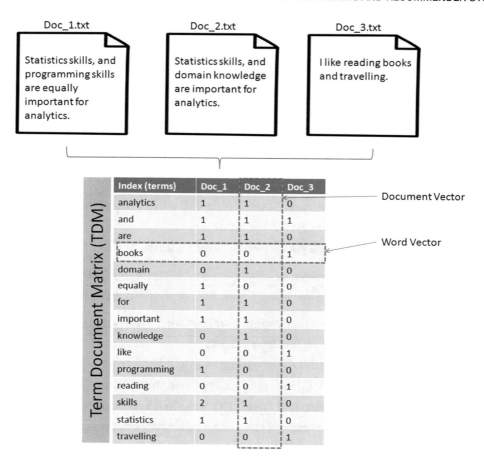

Figure 5-4. *Term Document Matrix*

Note   The Term Document Matrix (TDM) is the transpose of the term-document matrix. In TDM, the rows will be the document names and column headers will be the terms. Both are in the matrix format and useful for carrying out analysis; however, TDM is commonly used due to the fact that the number of terms tends to be usually much greater than the document count. In this case, having more rows is better than having a large number of columns.

Term Frequency-Inverse Document Frequency (TF-IDF)

In the area of information retrieval, TF-IDF is a good statistical measure to reflect the relevance of the term to the document in a collection of documents or corpus. Let's break down TF_IDF and apply an example to understand it better.

Term frequency will tell you how frequently a given term appears.

$$\text{TF (term)} = \frac{\textit{Number of times term appears in a document}}{\textit{Total number of terms in the document}}$$

For example, consider a document containing 100 words wherein the word "ML" appears three times, then TF (ML) = 3 / 100 = 0.03

Document frequency will tell you how important a term is.

$$\text{DF (term)} = \frac{d\left(\textit{number of documents containing a given term}\right)}{D\left(\textit{the size of the collection of documents}\right)}$$

Assume we have ten million documents and the word ML appears in one thousand of these, then DF (ML) = 1000 / 10,000,000 = 0.0001.

To normalize, let's take log (d/D), for example, log (0.0001) = -4

Quite often D > d and log (d/D) will give a negative value as seen in the preceding example. So to solve this problem, let's invert the ratio inside the log expression, which is known as inverse document frequency (IDF). Essentially we are compressing the scale of values so that very large or very small quantities are smoothly compared.

$$\text{IDF (term)} = \log\left(\frac{\textit{Total number of documents}}{\textit{Number of documents with a given term in it}}\right)$$

Continuing with the preceding example, IDF(ML) = log(10,000,000 / 1,000) = 4.

TF-IDF is the weight product of quantities; for the preceding example, TF-IDF (ML) = 0.03 * 4 = 0.12. Sklearn provides a function TfidfVectorizer to calculate TF-IDF for text; however, by default it normalizes the term vector using L2 normalization and also IDF is smoothed by adding 1 to the document frequency to prevent zero divisions (Listing 5-20).

Listing 5-20. Create a Term Document Matrix (TDM) with TF-IDF

```
from sklearn.feature_extraction.text import TfidfVectorizer

vectorizer = TfidfVectorizer()
doc_vec = vectorizer.fit_transform(docs.get('docs'))
#create dataFrame
df = pd.DataFrame(doc_vec.toarray().transpose(), index = vectorizer.get_
feature_names())

# Change column headers to be file names
df.columns = docs.get('ColNames')
print (df)
#----output----
```

	Doc_1.txt	Doc_2.txt	Doc_3.txt
analytics	0.276703	0.315269	0.000000
and	0.214884	0.244835	0.283217
are	0.276703	0.315269	0.000000
books	0.000000	0.000000	0.479528
domain	0.000000	0.414541	0.000000
equally	0.363831	0.000000	0.000000
for	0.276703	0.315269	0.000000
important	0.276703	0.315269	0.000000
knowledge	0.000000	0.414541	0.000000
like	0.000000	0.000000	0.479528
programming	0.363831	0.000000	0.000000
reading	0.000000	0.000000	0.479528
skills	0.553405	0.315269	0.000000
statistics	0.276703	0.315269	0.000000
travelling	0.000000	0.000000	0.479528

Data Exploration (Text)

In this stage the corpus is explored to understand the common key words, content, relationship, and presence and level of noise. This can be achieved by creating basic statistics and embracing visualization techniques such as word frequency count, word cooccurrence or correlation plot, etc., which will help us to discover hidden patterns, if any.

Frequency Chart

This visualization presents a bar chart whose length corresponds to the frequency a particular word occurred. Let's plot a frequency chart for Doc_1.txt file (Listing 5-21).

Listing 5-21. Example Code for Frequency Chart

```
words = df.index
freq = df.ix[:,0].sort(ascending=False, inplace=False)

pos = np.arange(len(words))
width=1.0
ax=plt.axes(frameon=True)
ax.set_xticks(pos)
ax.set_xticklabels(words, rotation='vertical', fontsize=9)
ax.set_title('Word Frequency Chart')
ax.set_xlabel('Words')
ax.set_ylabel('Frequency')
plt.bar(pos, freq, width, color='b')
plt.show()
#----output----
```

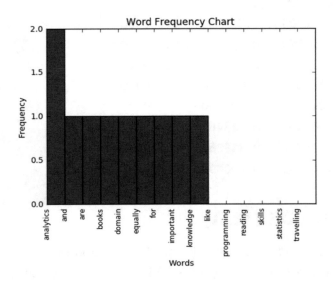

Word Cloud

This is a visual representation of text data, which is helpful to get a high-level understanding about the important keywords from data in terms of their occurrence. The wordcloud package can be used to generate words whose font size relates to their frequency (Listing 5-22).

Listing 5-22. Example Code for the Word Cloud

```
from wordcloud import WordCloud

# Read the whole text.
text = open('Data/Text_Files/Doc_1.txt').read()

# Generate a word cloud image
wordcloud = WordCloud().generate(text)

# Display the generated image:
# the matplotlib way:
import matplotlib.pyplot as plt
plt.imshow(wordcloud.recolor(random_state=2017))
plt.title('Most Frequent Words')
plt.axis("off")
plt.show()
#----output----
```

From the preceding chart we can see that "skills" appears the most number of times, comparatively.

Lexical Dispersion Plot

This plot is helpful to determine the location of a word in a sequence of text sentences. On the x-axis you'll have word offset numbers, and on the y-axis each row is a representation of entire text and the marker indicates an instance of the word of interest (Listing 5-23).

Listing 5-23. Example Code for Lexical Dispersion Plot

```
from nltk import word_tokenize

def dispersion_plot(text, words):
    words_token = word_tokenize(text)
    points = [(x,y) for x in range(len(words_token)) for y in
range(len(words)) if words_token[x] == words[y]]

    if points:
        x,y=zip(*points)
    else:
        x=y=()

    plt.plot(x,y,"rx",scalex=.1)
    plt.yticks(range(len(words)),words,color="b")
    plt.ylim(-1,len(words))
    plt.title("Lexical Dispersion Plot")
    plt.xlabel("Word Offset")
    plt.show()

text = 'statistics skills and programming skills are equally important
for analytics. statistics skills and domain knowledge are important for
analytics'
```

```
dispersion_plot(text, ['statistics', 'skills', 'and', 'important'])
#----output----
```

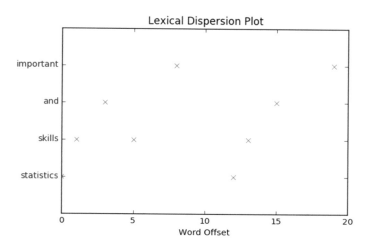

Cooccurrence Matrix

Calculating the cooccurrence between words in a sequence of text will be helpful matrices to explain the relationship between words. A cooccurrence matrix tells us how many times every word has cooccurred with the current word. Further plotting this matrix into a heat map is a powerful visual tool to spot the relationships between words efficiently (Listing 5-24).

Listing 5-24. Example Code for Cooccurrence Matrix

```
import statsmodels.api as sm
import scipy.sparse as sp

# default unigram model
count_model = CountVectorizer(ngram_range=(1,1))
docs_unigram = count_model.fit_transform(docs.get('docs'))

# co-occurrence matrix in sparse csr format
docs_unigram_matrix = (docs_unigram.T * docs_unigram)

# fill same word cooccurence to 0
docs_unigram_matrix.setdiag(0)
```

```
# co-occurrence matrix in sparse csr format
docs_unigram_matrix = (docs_unigram.T * docs_unigram) docs_unigram_matrix_
diags = sp.diags(1./docs_unigram_matrix.diagonal())

# normalized co-occurence matrix
docs_unigram_matrix_norm = docs_unigram_matrix_diags * docs_unigram_matrix

# Convert to a dataframe
df = pd.DataFrame(docs_unigram_matrix_norm.todense(), index = count_model.
get_feature_names())
df.columns = count_model.get_feature_names()

# Plot
sm.graphics.plot_corr(df, title='Co-occurrence Matrix', xnames=list(df.index))
plt.show()
#----output----
```

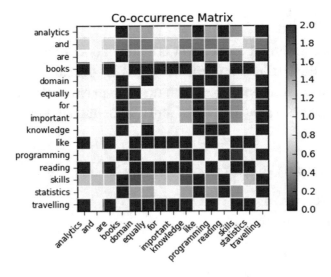

Model Building

As you might be familiar by now, model building is the process of understanding and establishing the relationship between variables. So far you have learned how to extract text content from various sources, preprocess to remove noise, and perform exploratory analysis to get basic understanding/statistics about the text data in hand. Now you'll learn to apply ML techniques on the processed data to build models.

Text Similarity

This is a measure that indicates how similar two objects are, described through a distance measure with dimensions represented by features of the objects (here, text). Smaller distance indicates a high degree of similarity and vice versa. Note that similarity is highly subjective and dependent on domain or application. For text similarity, it is important to choose the right distance measure to get better results. There are various distance measures available, with Euclidian metric being the most common, which is a straight line distance between two points. However a significant amount of research has been carried out in the field of text mining to learn that cosine distance is better suited for text similarity.

Let's look at a simple example (Table 5-4) to understand similarity better. Consider three documents containing certain simple text keywords and assume that the top two keywords are "accident" and "New York." For the moment, ignore other keywords and let's calculate the similarity of the document based on these two keywords' frequency.

Table 5-4. *Sample Term Document Matrix*

Document #	Count of 'Accident'	Count of 'New York'
1	2	8
2	3	7
3	7	3

Plotting the document word vector points on a two-dimensional chart is depicted in Figure 5-5. Notice that the cosine similarity equation is the representation of the angle between the two data points, whereas Euclidian distance is the square root of straight line difference between data points. The cosine similarity equation will result in a value between 0 and 1. The smaller cosine angle results in a bigger cosine value, indicating higher similarity. In this case, Euclidean distance will result in a zero. Let's put the values in the formula to find the similarity between documents 1 and 2.

$$\text{Euclidian distance (doc1, doc2)} = \sqrt{(2-3)^\wedge 2 + (8-7)^\wedge 2} = \sqrt{(1+1)} = 1.41 = 0$$

$$\text{Cosine (doc1, doc2)} = \frac{62}{8.24 * 7.61} = 0.98, \text{ where}$$

doc1 = (2,8)

doc2 = (3,7)

doc1 . doc2 = (2*3 + 8*7) = (56 + 6) = 62

$\|doc1\| = ` \sqrt{(2*2)+(8*8)} = 8.24$

$\|doc2\| = \sqrt{(3*3)+(7*7)} = 7.61$

Similarly let's find the similarity between document 1 and 3 (Figure 5-5).

Euclidian distance (doc1, doc3) = $\sqrt{(2-7)^2+(8-3)^2} = \sqrt{(25+25)} = 7.07 = 0$

Cosine (doc1, doc3)= $\dfrac{38}{8.24*7.61} = 0.60$

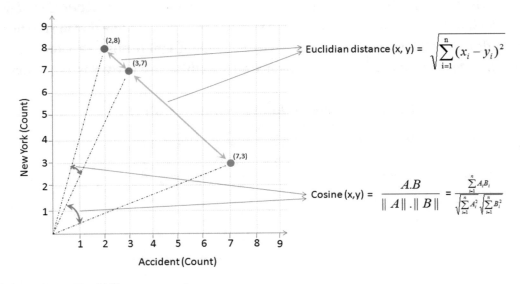

Figure 5-5. *Euclidian vs. cosine*

According to the cosine equation, documents 1 and 2 are 98% similar; this could mean that these two documents are talking more about New York, whereas document 3 can be assumed to be focused more about "Accident." However, there is a mention of New York a couple of times, resulting in a similarity of 60% between documents 1 and 3.

Listing 5-25 provides an example code for calculating cosine similarity for the example given in Figure 5-5.

Listing 5-25. Example Code for Calculating Cosine Similarity for Documents

```
from sklearn.metrics.pairwise import cosine_similarity

print "Similarity b/w doc 1 & 2: ", cosine_similarity([df['Doc_1.txt']],
[df['Doc_2.txt']])
print "Similarity b/w doc 1 & 3: ", cosine_similarity([df['Doc_1.txt']],
[df['Doc_3.txt']])
print "Similarity b/w doc 2 & 3: ", cosine_similarity([df['Doc_2.txt']],
[df['Doc_3.txt']])
#----output----
Similarity b/w doc 1 & 2:  [[ 0.76980036]]
Similarity b/w doc 1 & 3:  [[ 0.12909944]]
Similarity b/w doc 2 & 3:  [[ 0.1490712]]
```

Text Clustering

As an example, we'll be using the 20 newsgroups dataset, which consists of 18,000+ newsgroup posts on 20 topics. You can learn more about the dataset at http://qwone. com/~jason/20Newsgroups/. Let's load the data and check the topic names (Listing 5-26).

Listing 5-26. Example Code for Text Clustering

```
from sklearn.datasets import fetch_20newsgroups
from sklearn.feature_extraction.text import TfidfVectorizer
from sklearn.preprocessing import Normalizer
from sklearn import metrics
from sklearn.cluster import KMeans, MiniBatchKMeans
import numpy as np

# load data and print topic names
newsgroups_train = fetch_20newsgroups(subset='train')
print(list(newsgroups_train.target_names))
#----output----
['alt.atheism', 'comp.graphics', 'comp.os.ms-windows.misc', 'comp.sys.ibm.
pc.hardware', 'comp.sys.mac.hardware', 'comp.windows.x', 'misc.forsale',
'rec.autos', 'rec.motorcycles', 'rec.sport.baseball', 'rec.sport.hockey',
```

'sci.crypt', 'sci.electronics', 'sci.med', 'sci.space', 'soc.religion.
christian', 'talk.politics.guns', 'talk.politics.mideast', 'talk.politics.
misc', 'talk.religion.misc']

To keep it simple, let's filter only three topics. Assume that we do not know the topics.
Let's run the clustering algorithm and examine the keywords of each cluster.

```
categories = ['alt.atheism', 'comp.graphics', 'rec.motorcycles']

dataset = fetch_20newsgroups(subset='all', categories=categories,
shuffle=True, random_state=2017)

print("%d documents" % len(dataset.data))
print("%d categories" % len(dataset.target_names))

labels = dataset.target

print("Extracting features from the dataset using a sparse vectorizer")
vectorizer = TfidfVectorizer(stop_words='english')
X = vectorizer.fit_transform(dataset.data)
print("n_samples: %d, n_features: %d" % X.shape)
#----output----
2768 documents
3 categories
Extracting features from the dataset using a sparse vectorizer
n_samples: 2768, n_features: 35311
```

Latent Semantic Analysis (LSA)

LSA is a mathematical method that tries to bring out latent relationships within a
collection of documents. Rather than looking at each document isolated from the
others, it looks at all the documents as a whole and the terms within them to identify
relationships. Let's perform LSA by running Singular Value Decomposition (SVD) on the
data to reduce the dimensionality.

SVD of matrix $A = U * \sum * V^T$

r = rank of matrix X

U = column orthonormal $m * r$ matrix

$\sum$ = diagonal $r * r$ matrix with singular value sorted in descending order

V = column orthonormal $r * n$ matrix

In our case, we have three topics, 2,768 documents, and a 35,311 word vocabulary (Figure 5-6).

* Original matrix = $2768*35311 \sim 10^8$
* SVD = $3*2768 + 3 + 3*35311 \sim 10^{5.3}$

The resultant SVD takes up approximately 460 times less space than the original matrix. Listing 5-27 provides an example code for LSA through SVD.

Note   Latent semantic analysis (LSA) and latent semantic indexing (LSI) are the same thing, with the latter name being used sometimes when referring specifically to indexing a collection of documents for search (information retrieval).

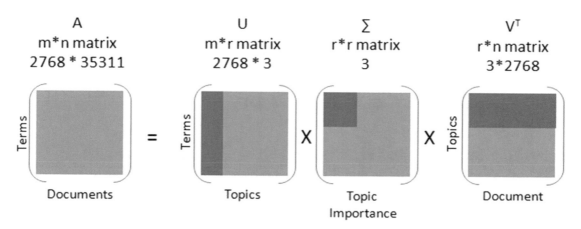

Figure 5-6. *Singular value decomposition*

Listing 5-27. Example Code for LSA Through SVD

```
from sklearn.decomposition import TruncatedSVD

# Lets reduce the dimensionality to 2000
svd = TruncatedSVD(2000)
lsa = make_pipeline(svd, Normalizer(copy=False))

X = lsa.fit_transform(X)
```

```
explained_variance = svd.explained_variance_ratio_.sum()
print("Explained variance of the SVD step: {}%".format(int(explained_
variance * 100)))
#----output----
Explained variance of the SVD step: 95%
```

Listing 5-28 is an example code to run k-means clustering on the SVD output.

Listing 5-28. k-means Clustering on SVD Dataset

```
from __future__ import print_function

km = KMeans(n_clusters=3, init='k-means++', max_iter=100, n_init=1)

# Scikit learn provides MiniBatchKMeans to run k-means in batch mode
suitable for a very large corpus
# km = MiniBatchKMeans(n_clusters=5, init='k-means++', n_init=1,
init_size=1000, batch_size=1000)

print("Clustering sparse data with %s" % km)
km.fit(X)

print("Top terms per cluster:")
original_space_centroids = svd.inverse_transform(km.cluster_centers_)
order_centroids = original_space_centroids.argsort()[:, ::-1]

terms = vectorizer.get_feature_names()
for i in range(3):
    print("Cluster %d:" % i, end=")
    for ind in order_centroids[i, :10]:
        print(' %s' % terms[ind], end=")
    print()
#----output----
Top terms per cluster:
Cluster 0: edu graphics university god subject lines organization com
posting uk
Cluster 1: com bike edu dod ca writes article sun like organization
Cluster 2: keith sgi livesey caltech com solntze wpd jon edu sandvik
```

Listing 5-29 is an example code to run hierarchical clustering on the SVD dataset.

Listing 5-29. Hierarchical Clustering on SVD Dataset

```
from sklearn.metrics.pairwise import cosine_similarity
dist = 1 - cosine_similarity(X)

from scipy.cluster.hierarchy import ward, dendrogram

linkage_matrix = ward(dist) #define the linkage_matrix using ward
clustering pre-computed distances

fig, ax = plt.subplots(figsize=(8, 8)) # set size
ax = dendrogram(linkage_matrix, orientation="right")

plt.tick_params(axis= 'x', which='both')

plt.tight_layout() #show plot with tight layout
plt.show()
#----output----
```

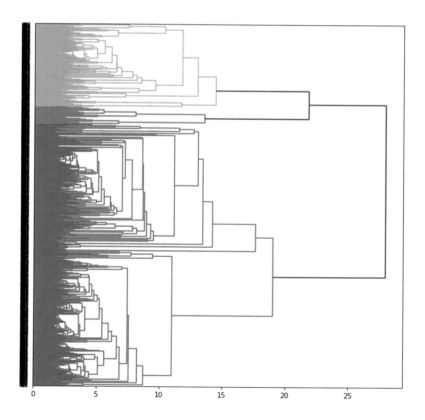

Topic Modeling

Topic modeling algorithms enable you to discover hidden topical patterns or thematic structure in a large collection of documents. The most popular topic modeling techniques are LDA and NMF.

Latent Dirichlet Allocation

LDA was presented by David Blei, Andrew Ng, and Michael I. Jordan in 2003 as a graphical model (Figure 5-7).

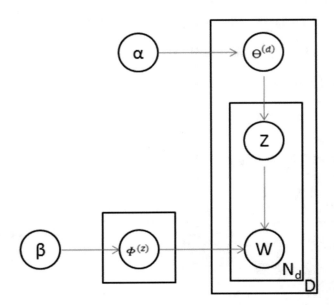

Figure 5-7. *LDA graph model*

LDA is given by $P(d, w) = P(d) * P(\Theta^{(d)}|\alpha) * \sum_{z} P(\Phi^{(z)}|\beta) * P(w|z, \Phi^{(z)}) * P(z|\Theta^{(d)})$

Where, $\Phi^{(z)}$ = word distribution for topic,

α = Dirichlet parameter prior the per-document topic distribution,

β = Dirichlet parameter prior the per-document word distribution,

$\Theta^{(d)}$ = topic distribution for a document.

LDA's objective is to maximize separation between means of projected topics and minimize variance within each projected topic. So LDA defines each topic as a bag of words by carrying out three steps described as follows (Figure 5-8).

Step 1: Initialize k clusters and assign each word in each document to one of the k topics.

Step 2: Reassign the word to a new topic based on a) the proportion of words for a document to a topic and b) the proportion of a topic widespread across all documents.

Step 3: Repeat step 2 until coherent topics result.

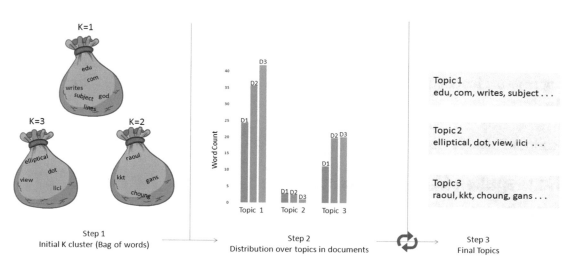

Figure 5-8. *Latent Dirichlet allocation (LDA)*

Listing 5-30 provides an example code for implementing LDA.

Listing 5-30. Example Code for LDA

```
from sklearn.decomposition import LatentDirichletAllocation

# continuing with the 20 newsgroup dataset and 3 topics
total_topics = 3
lda = LatentDirichletAllocation(n_components=total_topics,
                                max_iter=100,
                                learning_method='online',
                                learning_offset=50.,
                                random_state=2017)
lda.fit(X)

feature_names = np.array(vectorizer.get_feature_names())
```

```
for topic_idx, topic in enumerate(lda.components_):
    print("Topic #%d:" % topic_idx)
    print(" ".join([feature_names[i] for i in topic.argsort()[:-20 - 1:-1]]))
#----output----
Topic #0:
edu com writes subject lines organization article posting university nntp
host don like god uk ca just bike know graphics
Topic #1:
anl elliptical maier michael_maier qmgate separations imagesetter 5298 unscene
appreshed linotronic l300 iici amnesia glued veiw halftone 708 252 dot
Topic #2:
hl7204 eehp22 raoul vrrend386 qedbbs choung qed daruwala ims kkt briarcliff
kiat philabs col op_rows op_cols keeve 9327 lakewood gans
```

Nonnegative Matrix Factorization

NMF is a decomposition method for multivariate data and is given by V = MH, where V is the product of matrices W and H. W is a matrix of words rank in the features and H is the coefficient matrix, with each row being a feature. The three matrices have no negative elements (Listing 5-31).

Listing 5-31. Example Code for Nonnegative Matrix Factorization

```
from sklear.n.decomposition import NMF

nmf = NMF(n_components=total_topics, random_state=2017, alpha=.1, l1_ratio=.5)
nmf.fit(X)

for topic_idx, topic in enumerate(nmf.components_):
    print("Topic #%d:" % topic_idx)
    print(" ".join([feature_names[i] for i in topic.argsort()[:-20 - 1:-1]]))
#----output----
Topic #0:
edu com god writes article don subject lines organization just university
bike people posting like know uk ca think host
```

Topic #1:

sgi livesey keith solntze wpd jon caltech morality schneider cco moral com allan edu objective political cruel atheists gap writes

Topic #2:

sun east green ed egreen com cruncher microsystems ninjaite 8302 460 rtp 0111 nc 919 grateful drinking pixel biker showed

Text Classification

The ability to represent text features as numbers opens up the opportunity to run classification ML algorithms. Let's use a subset of 20 newsgroups data to build a classification model and assess its accuracy (Listing 5-32).

Listing 5-32. Example Code Text Classification on 20 News Groups Dataset

```
categories = ['alt.atheism', 'comp.graphics', 'rec.motorcycles', 'sci.
space', 'talk.politics.guns']

newsgroups_train = fetch_20newsgroups(subset='train',
categories=categories, shuffle=True, random_state=2017, remove=('headers',
'footers', 'quotes'))
newsgroups_test = fetch_20newsgroups(subset='test', categories=categories,
shuffle=True, random_state=2017, remove=('headers', 'footers', 'quotes'))

y_train = newsgroups_train.target
y_test = newsgroups_test.target

vectorizer = TfidfVectorizer(sublinear_tf=True, smooth_idf = True, max_
df=0.5,  ngram_range=(1, 2), stop_words='english')
X_train = vectorizer.fit_transform(newsgroups_train.data)
X_test = vectorizer.transform(newsgroups_test.data)

print("Train Dataset")
print("%d documents" % len(newsgroups_train.data))
print("%d categories" % len(newsgroups_train.target_names))
print("n_samples: %d, n_features: %d" % X_train.shape)
```

```
print("Test Dataset")
print("%d documents" % len(newsgroups_test.data))
print("%d categories" % len(newsgroups_test.target_names))
print("n_samples: %d, n_features: %d" % X_test.shape)
#----output----
Train Dataset
2801 documents
5 categories
n_samples: 2801, n_features: 241036
Test Dataset
1864 documents
5 categories
n_samples: 1864, n_features: 241036
```

Let's build a simple naïve Bayes classification model and assess the accuracy. Essentially we can replace naïve Bayes with any other classification algorithm or use an ensemble model to build an efficient model (Listing 5-33).

Listing 5-33. Example Code Text Classification Using Multinomial Naïve Bayes

```
from sklearn.naive_bayes import MultinomialNB
from sklearn import metrics

clf = MultinomialNB()
clf = clf.fit(X_train, y_train)

y_train_pred = clf.predict(X_train)
y_test_pred = clf.predict(X_test)

print ('Train accuracy_score: ', metrics.accuracy_score(y_train, y_train_
pred))
print ('Test accuracy_score: ',metrics.accuracy_score(newsgroups_test.
target, y_test_pred))

print ("Train Metrics: ", metrics.classification_report(y_train, y_train_
pred))
print ("Test Metrics: ", metrics.classification_report(newsgroups_test.
target, y_test_pred))
#----output----
```

```
Train accuracy_score:  0.9760799714387719
Test accuracy_score:  0.8320815450643777
```

Train Metrics:

	precision	recall	f1-score	support
0	1.00	0.97	0.98	480
1	1.00	0.97	0.98	584
2	0.91	1.00	0.95	598
3	0.99	0.97	0.98	593
4	1.00	0.97	0.99	546
micro avg	0.98	0.98	0.98	2801
macro avg	0.98	0.98	0.98	2801
weighted avg	0.98	0.98	0.98	2801

Test Metrics:

	precision	recall	f1-score	support
0	0.91	0.62	0.74	319
1	0.90	0.90	0.90	389
2	0.81	0.90	0.86	398
3	0.80	0.84	0.82	394
4	0.78	0.86	0.82	364
micro avg	0.83	0.83	0.83	1864
macro avg	0.84	0.82	0.83	1864
weighted avg	0.84	0.83	0.83	1864

Sentiment Analysis

The procedure of discovering and classifying opinions expressed in a piece of text (like comments/feedback text) is called the sentiment analysis. The intended output of this analysis would be to determine whether the writer's mindset toward a topic, product, service, etc. is neutral, positive, or negative (Listing 5-34).

Listing 5-34. Example Code for Sentiment Analysis

```
from nltk.sentiment.vader import SentimentIntensityAnalyzer
from nltk.sentiment.util import *
data = pd.read_csv('Data/customer_review.csv')
```

```
SIA = SentimentIntensityAnalyzer()
data['polarity_score']=data.Review.apply(lambda x:SIA.polarity_scores(x)
['compound'])
data['neutral_score']=data.Review.apply(lambda x:SIA.polarity_scores(x)
['neu'])
data['negative_score']=data.Review.apply(lambda x:SIA.polarity_scores(x)
['neg'])
data['positive_score']=data.Review.apply(lambda x:SIA.polarity_scores(x)
['pos'])
data['sentiment']="
data.loc[data.polarity_score>0,'sentiment']='POSITIVE'
data.loc[data.polarity_score==0,'sentiment']='NEUTRAL'
data.loc[data.polarity_score<0,'sentiment']='NEGATIVE'
data.head()

data.sentiment.value_counts().plot(kind='bar',title="sentiment analysis")
plt.show()
#----output----
```

	ID	Review	polarity_score
0	1	Excellent service my claim was dealt with very...	0.7346
1	2	Very sympathetically dealt within all aspects ...	-0.8155
2	3	Having received yet another ludicrous quote fr...	0.9785
3	4	Very prompt and fair handling of claim. A mino...	0.1440
4	5	Very good and excellent value for money simple...	0.8610

	neutral_score	negative_score	positive_score	sentiment
0	0.618	0.000	0.382	POSITIVE
1	0.680	0.320	0.000	NEGATIVE
2	0.711	0.039	0.251	POSITIVE
3	0.651	0.135	0.214	POSITIVE
4	0.485	0.000	0.515	POSITIVE

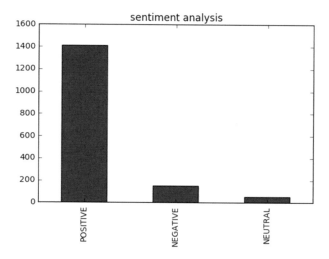

Deep Natural Language Processing (DNLP)

First, let me clarify that DNLP is not to be mistaken for deep learning NLP. A technique such as topic modeling is generally known as shallow NLP where you try to extract knowledge from text through semantic or syntactic analysis approach (i.e., try to form groups by retaining words that are similar and hold higher weight in a sentence/document). Shallow NLP is less noisy than the n-grams; however, the key drawback is that it does not specify the role of items in the sentence. In contrast, DNLP focuses on a semantic approach. That is, it detects relationships within the sentences, and further, it can be represented or expressed as a complex construction of the form such as subject:predicate:object (known as triples or triplets) out of syntactically parsed sentences to retain the context. Sentences are made up of any combination of actor, action, object, and named entities (persons, organizations, locations, dates, etc.). For example, consider the sentence "the flat tire was replaced by the driver." Here, "driver" is the subject (actor), "replaced" is the predicate (action), and "flat tire" is the object. So the triples for that would be the driver:replaced:tire, which captures the context of the sentence. Note that triples are one of the forms widely used, and you can form similar complex structures based on the domain or problem at hand.

For ae demonstration, I'll use the sopex package, which uses the Stanford Core NLP tree parser (Listing 5-35).

Listing 5-35. Example Code for Deep NLP

```
from chunker import PennTreebackChunker
from extractor import SOPExtractor

# Initialize chunker
chunker = PennTreebackChunker()
extractor = SOPExtractor(chunker)

# function to extract triples
def extract(sentence):
    sentence = sentence if sentence[-1] == '.' else sentence+'.'
    global extractor
    sop_triplet = extractor.extract(sentence)
    return sop_triplet

sentences = [
  'The quick brown fox jumps over the lazy dog.',
  'A rare black squirrel has become a regular visitor to a suburban garden',
  'The driver did not change the flat tire',
  "The driver crashed the bike white bumper"
]

#Loop over sentence and extract triples
for sentence in sentences:
    sop_triplet = extract(sentence)
    print sop_triplet.subject + ':' + sop_triplet.predicate + ':' + sop_
    triplet.object
#----output----
fox:jumps:dog
squirrel:become:visitor
driver:change:tire
driver:crashed:bumper
```

Word2Vec

The Tomas Mikolov led team at Google created Word2Vec (word to vector) model in 2013, which uses documents to train a neural network model to maximize the conditional probability of context, given the word.

It utilizes two models: CBOW and skip-gram.

1. The continuous bag-of-words (CBOW) model predicts the current word from a window of surrounding context words or, given a set of context words, predicts the missing word that is likely to appear in that context. CBOW is faster than skip-gram to train and gives better accuracy for frequently appearing words.

2. The continuous skip-gram model predicts the surrounding window of context words using the current word or, given a single word, predicts the probability of other words that are likely to appear near it in that context. Skip-gram is known to show good results for both frequent and rare words. Let's look at an example sentence and create a skip-gram for a window of 2 (Figure 5-9). The word highlighted in yellow is the input word.

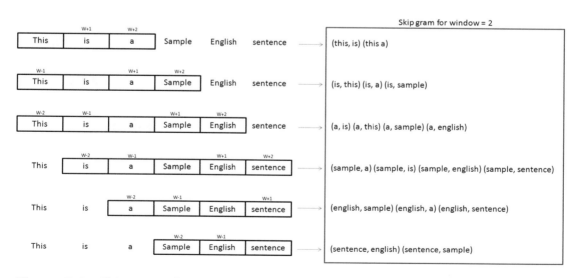

Figure 5-9. *Skip-gram for window of 2*

You can download Google's pretrained model (from the following link) for Word2Vec, which includes the vocabulary of 3 million words/phrases taken from 100 billion words from a Google News dataset.

URL: `https://drive.google.com/file/d/0B7XkCwpI5KDYNlNUTTlSS21pQmM/edit`

Listing 5-36 provides an example code for Word2Vec implementation.

Listing 5-36. Example Code for Word2Vec

```
import gensim

# Load Google's pre-trained Word2Vec model.
model = gensim.models. KeyedVectors.load_word2vec_format('Data/GoogleNews-
vectors-negative300.bin', binary=True)

model.most_similar(positive=['woman', 'king'], negative=['man'], topn=5)
#----output----
[(u'queen', 0.7118192911148071),
 (u'monarch', 0.6189674139022827),
 (u'princess', 0.5902431607246399),
 (u'crown_prince', 0.5499460697174072),
 (u'prince', 0.5377321243286133)]

model.most_similar(['girl', 'father'], ['boy'], topn=3)
#----output----
[(u'mother', 0.831214427947998),
 (u'daughter', 0.8000643253326416),
 (u'husband', 0.769158124923706)]

model.doesnt_match("breakfast cereal dinner lunch".split())
#----output----
'cereal'
```

You can train a Word2Vec model on your own data set. The key model parameters to be remembered are size, window, min_count, and sg (Listing 5-37).

size: The dimensionality of the vectors. The bigger size values require more training data but can lead to more accurate models.

sg = 0 for CBOW model and 1 for skip-gram model.

min_count: Ignore all words with a total frequency lower than this.

window: The maximum distance between the current and predicted word within a sentence.

Listing 5-37. Example Code for Training word2vec on Your Own Dataset

```
sentences = [['cigarette','smoking','is','injurious', 'to', 'health'],
['cigarette','smoking','causes','cancer'],['cigarette','are','not',
'to','be','sold','to','kids']]
```

```
# train word2vec on the two sentences
model = gensim.models.Word2Vec(sentences, min_count=1, sg=1, window = 3)
```

```
model.most_similar(positive=['cigarette', 'smoking'], negative=['kids'],
topn=1)
#----output----
[('injurious', 0.16142114996910095)]
```

Recommender Systems

Personalization of the user experience has been a high priority and has become the new mantra in consumer-focused industries. You might have observed e-commerce companies casting personalized ads for you suggesting what to buy, which news to read, which video to watch, where/what to eat, and who you might be interested in networking with (friends/professionals) on social media sites. Recommender systems are the core information filtering system designed to predict user preference and help to recommend the right items to create a user-specific personalization experience. There are two types of recommendation systems: 1) content-based filtering and 2) collaborative filtering (Figure 5-10).

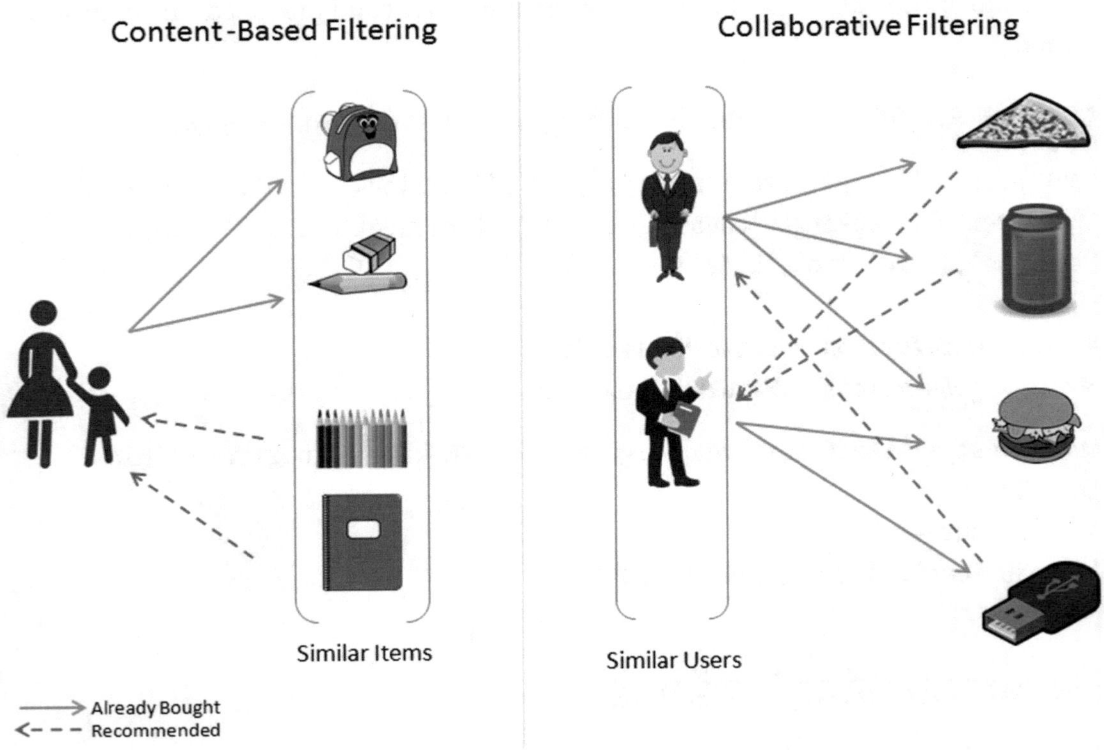

Figure 5-10. *Recommender systems*

Content-Based Filtering

This type of system focuses on the similarity attribute of the items to give you recommendations. This is best understood with an example: if a user has purchased items of a particular category, other similar items from the same category are recommended to the user (refer to Figure 5-10).

The item-based similarity recommendation algorithm can be represented as:

$$\hat{x}_{k,m} = \frac{\sum_{i_b} sim_i\left(i_m, i_b\right)\left(x_{k,b}\right)}{\sum_{i_b}\left|sim_i\left(i_m, i_b\right)\right|}$$

Collaborative Filtering (CF)

CF focuses on the similarity attribute of the users, that is, it finds people with similar tastes based on a similarity measure from the large group of users. There are two types of CF implementation in practice: memory-based and model-based.

The memory-based type is mainly based on the similarity algorithm; the algorithm looks at items liked by similar people to create a ranked list of recommendations. You can then sort the ranked list to recommend the top n items to the user.

The user-based similarity recommendation algorithm can be represented as:

$$pr_{x,k} = m_x + \frac{\sum_{u_y \in N_x} \left(r_{y,k} - m_y \right) \mathrm{sim}\left(\boldsymbol{u}_x, \boldsymbol{u}_y \right)}{\sum_{u_y \in N_x} \left| \mathrm{sim}\left(\boldsymbol{u}_x, \boldsymbol{u}_y \right) \right|}$$

Let's consider an example data set of movie rating (Figure 5-11) and apply item- and user-based recommendations to get a better understanding. Listing 5-38 provides an example code for a recommender system.

User ID	Anaconda (Thriller/Adventure)	Avengers Assemble (Fantasy/Science)	A Walk to Remember (Romance)	Avatar (Fantasy/Science)	50 First Dates (Romance)	Interstellar (Fantasy/Science)
1	2.5	3.5	3	3.5	2.5	3
2	3	3.5	1.5	5	3.5	3
3	2.5	3	0	3.5	0	4
4	0	3.5	3	4	2.5	4.5
5	3	4	2	3	2	3
6	3	4	0	5	3.5	3
7	4.5	0	0	4	1	0

Figure 5-11. *Movie rating sample data set*

Listing 5-38. Example Code for a Recommender System

```
import numpy as np
import pandas as pd

df = pd.read_csv('Data/movie_rating.csv')

n_users = df.userID.unique().shape[0]
n_items = df.itemID.unique().shape[0]
print ('\nNumber of users = ' + str(n_users) + ' | Number of movies = ' +
str(n_items))
```

```
#----output----
Number of users = 7 | Number of movies = 6

# Create user-item similarity matrices
df_matrix = np.zeros((n_users, n_items))
for line in df.itertuples():
    df_matrix[line[1]-1, line[2]-1] = line[3]

from sklearn.metrics.pairwise import pairwise_distances

user_similarity = pairwise_distances(df_matrix, metric='cosine')
item_similarity = pairwise_distances(df_matrix.T, metric='cosine')

# Top 3 similar users for user id 7
print ("Similar users for user id 7: \n", pd.DataFrame(user_
similarity).loc[6,pd.DataFrame(user_similarity).loc[6,:] > 0].sort_
values(ascending=False)[0:3])
#----output----
Similar users for user id 7:
3    8.000000
0    6.062178
5    5.873670

# Top 3 similar items for item id 6
print ("Similar items for item id 6: \n", pd.DataFrame(item_
similarity).loc[5,pd.DataFrame(item_similarity).loc[5,:] > 0].sort_
values(ascending=False)[0:3])
#----output----
0    6.557439
2    5.522681
3    4.974937
```

Let's build the user-based prediction and item-based prediction formula as a function. Apply this function to predict ratings and use root mean squared error (RMSE) to evaluate the model performance (Listing 5-39).

Listing 5-39. Example Code for Recommender System and Accuracy Evaluation

```
# Function for item based rating prediction
def item_based_prediction(rating_matrix, similarity_matrix):
    return rating_matrix.dot(similarity_matrix) / np.array([np.
    abs(similarity_matrix).sum(axis=1)])

# Function for user based rating prediction
def user_based_prediction(rating_matrix, similarity_matrix):
    mean_user_rating = rating_matrix.mean(axis=1)
    ratings_diff = (rating_matrix - mean_user_rating[:, np.newaxis])
    return mean_user_rating[:, np.newaxis] + similarity_matrix.dot(ratings_
    diff) / np.array([np.abs(similarity_matrix).sum(axis=1)]).T

item_based_prediction = item_based_prediction(df_matrix, item_similarity)
user_based_prediction = user_based_prediction(df_matrix, user_similarity)

# Calculate the RMSE
from sklearn.metrics import mean_squared_error
from math import sqrt
def rmse(prediction, actual):
    prediction = prediction[actual.nonzero()].flatten()
    actual = actual[actual.nonzero()].flatten()
    return sqrt(mean_squared_error(prediction, actual))

print ('User-based CF RMSE: ' + str(rmse(user_based_prediction, df_
matrix)))
print ('Item-based CF RMSE: ' + str(rmse(item_based_prediction, df_
matrix)))
#----output----
User-based CF RMSE: 1.0705767849
Item-based CF RMSE: 1.37392288971

y_user_based = pd.DataFrame(user_based_prediction)

# Predictions for movies that the user 6 hasn't rated yet
predictions = y_user_based.loc[6,pd.DataFrame(df_matrix).loc[6,:] == 0]
top = predictions.sort_values(ascending=False).head(n=1)
recommendations = pd.DataFrame(data=top)
```

```
recommendations.columns = ['Predicted Rating']
print (recommendations)
#----output----
   Predicted Rating
1          2.282415
```

```
y_item_based = pd.DataFrame(item_based_prediction)
```

```
# Predictions for movies that the user 6 hasn't rated yet
predictions = y_item_based.loc[6,pd.DataFrame(df_matrix).loc[6,:] == 0]
top = predictions.sort_values(ascending=False).head(n=1)
recommendations = pd.DataFrame(data=top)
recommendations.columns = ['Predicted Rating']
print (recommendations)
#----output----
   Predicted Rating
5          2.262497
```

Per user-based, the movie "Interstellar" is recommended (index number 5).
Per item-based, the recommended movie is Avengers Assemble (index number 1).

Model-based CF is based on matrix factorization (MF) such as SVD and NMF, etc.
Let's look at how to implement using SVD (Listing 5-40).

Listing 5-40. Example Code for Recommender System Using SVD

```
# calculate sparsity level
sparsity=round(1.0-len(df)/float(n_users*n_items),3)
print 'The sparsity level of is ' +  str(sparsity*100) + '%'
```

```
import scipy.sparse as sp
from scipy.sparse.linalg import svds
```

```
# Get SVD components from train matrix. Choose k.
u, s, vt = svds(df_matrix, k = 5)
s_diag_matrix=np.diag(s)
X_pred = np.dot(np.dot(u, s_diag_matrix), vt)
print ('User-based CF MSE: ' + str(rmse(X_pred, df_matrix)))
```

```
#----output----
The sparsity level of is 0.0%
User-based CF MSE: 0.015742898995
```

Note that in our case the data set is small, hence the sparsity level is 0%. I recommend you to try this method on the MovieLens 100k dataset, which you can download from `https://grouplens.org/datasets/movielens/100k/`.

Summary

In this step, you learned the fundamentals of the text mining process, and different tools/techniques to extract text from various file formats. You also learned the basic text preprocessing steps to remove noise from data, and different visualization techniques to get a better understanding of the corpus at hand. Then you learned various models that can be built to understand the relationships and get insight out of the data.

We also learned two important recommender system methods such as content-based filtering and collaborative filtering.

CHAPTER 6

Step 6: Deep and Reinforcement Learning

Deep learning has been the buzzword in the machine learning (ML) world in recent times. The main objective of deep learning algorithms so far has been to use ML to achieve artificial general intelligence (AGI) (i.e., replicate human level intelligence in machines to solve any problems for a given area). Deep learning has shown promising outcomes in computer vision, audio processing, and text mining. The advancements in this area led to breakthroughs such as self-driving cars. In this chapter, you'll learn about deep leaning's core concepts, evolution (perceptron to convolutional neural network [CNN]), key applications, and implementation.

There have been a number of powerful and popular open source libraries built in the last few years, predominantly focused on deep learning (Table 6-1).

Table 6-1. *Popular Deep Learning Libraries (as of the end of the year 2019)*

Library Name	Launch Year	License	# of Contributors	Official website
Theano	2010	BSD	333	http://deeplearning.net/software/theano/
Pylearn2	2011	BSD-3-Clause	115	http://deeplearning.net/software/pylearn2/
TensorFlow	2015	Apache-2.0	1963	http://tensorflow.org
PyTorch	2016	BSD	1023	https://pytorch.org/
Keras	2015	MIT	792	https://keras.io/

(*continued*)

© Manohar Swamynathan 2019
M. Swamynathan, *Mastering Machine Learning with Python in Six Steps,*
https://doi.org/10.1007/978-1-4842-4947-5_6

Table 6-1. (*continued*)

Library Name	Launch Year	License	# of Contributors	Official website
MXNet	2015	Apache-2.0	684	`http://mxnet.io/`
Caffe	2015	BSD-2-Clause	266	`http://caffe.` `berkeleyvision.org/`
Lasagne	2015	MIT	65	`http://lasagne.` `readthedocs.org/`

Following is a short description of each of the libraries (from Table 6-1). Their official websites provide quality documentation and examples. I strongly recommend that you visit the respective sites to learn more if required post completion of this chapter.

- *Thean*o: It is a Python library predominantly developed by academics at the Universite de Montreal. Theano allows you to define, optimize, and evaluate mathematical expressions involving complex multidimensional arrays efficiently. It is designed to work with GPUs and perform efficient symbolic differentiation. It is fast and stable, with an extensive unit test in place.

- *TensorFlow*: As per the official documentation, it is a library for numerical computation using data flow graphs for scalable ML, developed by Google researchers. It is currently being used by Google products for research and production. It was open sourced in 2015 and has gained wide popularity in the ML world.

- *Pylearn2*: An ML library based on Theano, which means users can write new models/algorithms using mathematical expressions and Theano will optimize, stabilize, and compile those expressions.

- *PyTorch*: It is an open source deep learning platform that provides a seamless path from research prototyping to production deployment. It has key features of hybrid front-end, distributed training, allows popular Python libraries to be used, and a rich ecosystem of tools/libraries extends PyTorch.

- *Keras*: It is known as a high-level neural networks library, written in Python and capable of running on top of either TensorFlow or Theano. It's an interface rather than an end-end ML framework. It's written in Python, simple to get started, highly modular, and easy yet deep enough to expand to build/support complex models.

- *MXNet*: It was developed in collaboration with researchers from CMU, NYU, NUS, and MIT. It's a lightweight, portable, flexible, distributed/mobile library support across many languages such as Python, R, Julia, Scala, Go, Javascript, etc.

- *Caffe*: It is a deep learning framework by Berkeley Vision and Learning Center written in C++, and has Python/Matlab-building capabilities.

- *Lasagne*: It is a lightweight library to build and train neural networks in Theano.

Throughout this chapter, Scikit-learn and Keras library with backend as TensorFlow or Theano have been used appropriately, due to the fact that these are the best choices for a beginner to get hold of the concepts. Also, these are the most widely used by ML practitioners.

Note   There is enough good material available on how to set up Keras with TensorFlow or Theano, so the same will not be covered here. Also, remember to install "graphviz" and "pydot-ng" packages to support the graphical view of a neural network. The Keras codes in this chapter were built on the Linux platform; however, they should work fine on other platforms without any modifications, providing that supporting packages are rightly installed.

Artificial Neural Network (ANN)

Before jumping into the details of deep learning, I think it is very important to briefly understand how human vision works. The human brain is a complex, connected neural network where different regions of the brain are responsible for different jobs; these regions are machines of the brain that receive signals and process them to take necessary action. Figure 6-1 shows the visual pathway of the human brain.

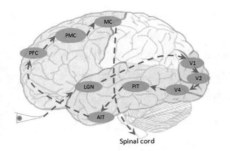

LGN - Lateral Geniculate Nucleus is the relay center for visual pathway
V1 - Primary Visual Cortex
V2 - Relays signals
V4 - Intermediate visual forms, color, feature groups etc
PIT - Posterior Inferior Temporal cortex
AIT - High level object description faces, objects
PFC - Decision making, categorical judgments
PMC - Premotor cortex - helps you control your movements
MC - Motor command

Figure 6-1. *Visual pathway*

Our brain is made up of a cluster of small connected units called neurons, which send electrical signals to one another. The long-term knowledge is represented by the strength of the connections between neurons. When we see objects, light travels through the retina and the visual information gets converted to electrical signals. Further, the electric signal passes through the hierarchy of connected neurons of different regions within the brain in a few milliseconds to decode signals/information.

What Goes On Behind, When Computers Look at an Image?

In computers, an image is represented as one large 3-dimensional array of numbers. For example, consider Figure 6-2: it is a handwritten grayscale digit image of 28×28×1 (width × height × depth) size resulting in 784 data points. Each number in the array is an integer that ranges from 0 (black) to 255 (white). In a typical classification problem, the model has to turn this large matrix into a single label. For a color image, additionally it will have three color channels—Red, Green, Blue (RGB) for each pixel—so the same image in color would be of size 28×28×3 = 2,352 data points.

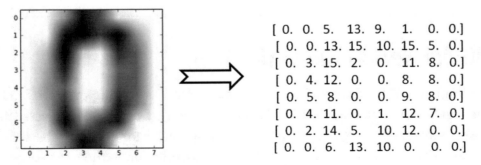

```
[ 0.  0.  5.  13.  9.   1.   0.  0.]
[ 0.  0.  13. 15.  10.  15.  5.  0.]
[ 0.  3.  15. 2.   0.   11.  8.  0.]
[ 0.  4.  12. 0.   0.   8.   8.  0.]
[ 0.  5.  8.  0.   0.   9.   8.  0.]
[ 0.  4.  11. 0.   1.   12.  7.  0.]
[ 0.  2.  14. 5.   10.  12.  0.  0.]
[ 0.  0.  6.  13.  10.  0.   0.  0.]
```

Figure 6-2. *Handwritten digit (zero) image and corresponding array*

Why Not a simple Classification Model for Images?

Image classification can be challenging for a computer, as there are a variety of challenges associated with the representation of the images. A simple classification model might not be able to address most of these issues without a lot of feature engineering effort. Let's understand some of the key issues (Table 6-2).

Table 6-2. *Visual Challenges in Image Data*

Description	Example
Viewpoint variation: The same object can have a different orientation.	
Scale and illumination variation: Variation in objects' size and the level of illumination on pixel level can vary.	
Deformation/twist and intraclass variation: Nonrigid bodies can be deformed in great ways, and there can be different types of objects with varying appearance within a class.	
Blockage: Only a small portion of the object of interest may be visible.	
Background clutter: Objects can blend into their environment, which will make them hard to identify.	

Perceptron—Single Artificial Neuron

Inspired by the biological neurons, McCulloch and Pitts in 1943 introduced the concept of perceptron as an artificial neuron, which is the basic building block of an ANN. They are not only named after their biological counterparts, but also modeled after the behavior of the neurons in our brain (Figure 6-3).

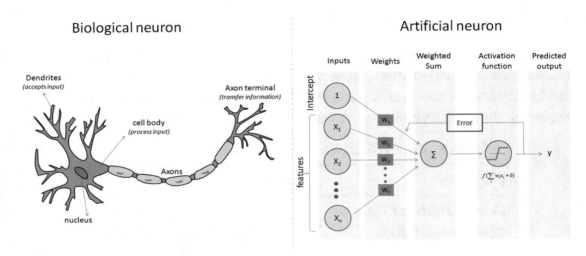

Figure 6-3. *Biological vs. artificial neuron*

A biological neuron has dendrites to receive signals, a cell body to process them, and an axon/axon terminal to transfer signals out to other neurons. Similarly, an artificial neuron has multiple input channels to accept training samples represented as a vector, and a processing stage where the weights(w) are adjusted such that the output error (actual vs. predicted) is minimized. Then the result is fed into an activation function to produce output, for example, a classification label. The activation function for a classification problem is a threshold cutoff (standard is .5) above which class is 1, else 0. Let's see how this can be implemented using Scikit-learn (Listing 6-1).

Listing 6-1. Example Code for Sklearn Perceptron

```
# import sklearn.linear_model.perceptron
from sklearn.linear_model import perceptron
import matplotlib.pyplot as plt
from matplotlib.colors import ListedColormap

# Let's use sklearn to make_classification function to create some test data.
from sklearn.datasets import make_classification
X, y = make_classification(20, 2, 2, 0, weights=[.5, .5], random_
state=2017)
```

```
# Create the model
clf = perceptron.Perceptron(n_iter=100, verbose=0, random_state=2017, fit_
intercept=True, eta0=0.002)
clf.fit(X,y)

# Print the results
print ("Prediction: " + str(clf.predict(X)))
print ("Actual:     " + str(y))
print ("Accuracy:   " + str(clf.score(X, y)*100) + "%")

# Output the values
print ("X1 Coefficient: " + str(clf.coef_[0,0]))
print ("X2 Coefficient: " + str(clf.coef_[0,1]))
print ("Intercept:      " + str(clf.intercept_))

# Plot the decision boundary using custom function 'plot_decision_regions'
plot_decision_regions(X, y, classifier=clf)
plt.title('Perceptron Model Decision Boundry')
plt.xlabel('X1')
plt.ylabel('X2')
plt.legend(loc='upper left')
plt.show()
#----output----
Prediction: [1 1 1 0 0 0 0 1 0 1 1 0 0 1 0 1 0 0 1 1]
Actual:     [1 1 1 0 0 0 0 1 0 1 1 0 0 1 0 1 0 0 1 1]
Accuracy:   100.0%
X1 Coefficient: 0.00575308754305
X2 Coefficient: 0.00107517941422
Intercept:      [-0.002]
```

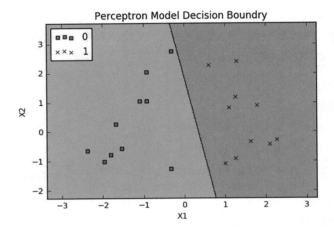

Note   The drawback of the single perceptron approach is that it can only learn linearly separable functions.

Multilayer Perceptrons (Feedforward Neural Network)

To address the drawback of the single perceptron, multilayer perceptrons was proposed, also commonly known as a feedforward neural network. It is a composition of multiple perceptrons connected in different ways and operating on distinctive activation functions, to enable improved learning mechanism. The training sample propagates forward through the network and the output error is back propagated; the error is minimized using a gradient descent method, which will calculate a loss function for all the weights in the network (Figure 6-4).

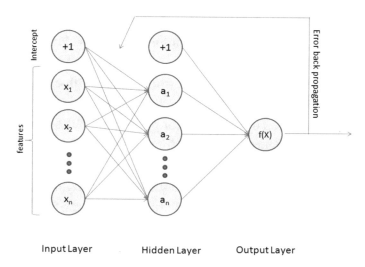

Figure 6-4. *Multilayer perceptron representation*

The activation function for a simple one-level hidden layer of a multilayer perceptron can be given by:

$$f(x) = g\left(\sum_{j=0}^{M} W_{kj}^{(2)} g\left(\sum_{i=0}^{d} W_{ji}^{(1)} x_i \right) \right),$$ where x_i is the input and $W_{ji}^{(1)}$ is the input layer

weights and $W_{kj}^{(2)}$ is the weight of the hidden layer.

The multilayered neural network can have many hidden layers, where the network holds its internal abstract representation of the training sample. The upper layers will be building new abstractions on top of the previous layers. So having more hidden layers for the complex dataset will help the neural network to learn better.

As you can see from Figure 6-4, the MLP (multilayer perceptron) architecture has a minimum of three layers: input, hidden, and output layers. The input layer's neuron count will be equal to the total number of features and in some libraries, an additional neuron for intercept/bias. These neurons are represented as nodes. The output layers will have a single neuron for regression models and binary classifier; otherwise it will be equal to the total number of class labels for multiclass classification models.

Note that using too few neurons for the complex dataset can result in an underfitted model, due to the fact that it might fail to learn the patterns in complex data. However, using too many neurons can result in an overfitted model, as it has the capacity to capture a pattern that might be noise or specific for the given training data set. So, to build an efficient multilayered neural network, the fundamental questions to be answered about hidden layers for implementation are: 1) what is the ideal number of hidden layers? 2) What should be the number of neurons in hidden layers?

The widely accepted rule of thumb is that you can start with one hidden layer, as there is a theory that one hidden layer is sufficient for the majority of problems. Then gradually increase the layers on a trial and error basis to see if there is any improvement in accuracy. The number of neurons in the hidden layer can ideally be the mean of the neurons in the input and output layers.

Let's see the MLP algorithm in action from the Scikit-learn library on a classification problem. We'll be using the digits dataset available as part of the Scikit-learn dataset, which is made up of 1,797 samples (a subset of the MNIST dataset)—handwritten grayscale digit 8×8 images.

Load MNIST Data

Listing 6-2 provides an example code for loading MNIST data for training MLPClassifier. MNIST digit data is part of the sklearn datasets.

Listing 6-2. Example Code for Loading MNIST Data for Training MLPClassifier

```
import pandas as pd
import numpy as np
import matplotlib.pyplot as plt

from sklearn.neural_network import MLPClassifier
from sklearn.model_selection import train_test_split
from sklearn.preprocessing import StandardScaler
from sklearn.metrics import confusion_matrix

from sklearn.datasets import load_digits
np.random.seed(seed=2017)

# load data
digits = load_digits()
print('We have %d samples'%len(digits.target))
```

```
## plot the first 32 samples to get a sense of the data
fig = plt.figure(figsize = (8,8))
fig.subplots_adjust(left=0, right=1, bottom=0, top=1, hspace=0.05,
wspace=0.05)
for i in range(32):
    ax = fig.add_subplot(8, 8, i+1, xticks=[], yticks=[])
    ax.imshow(digits.images[i], cmap=plt.cm.gray_r)
ax.text(0, 1, str(digits.target[i]), bbox=dict(facecolor='white'))
#----output----
We have 1797 samples
```

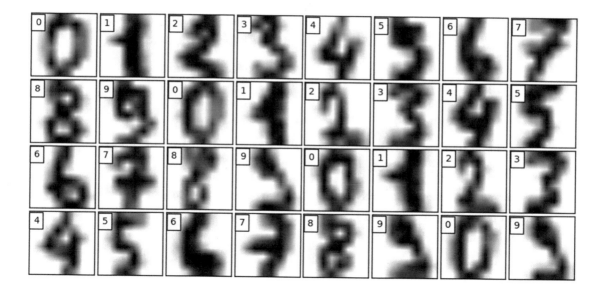

Key Parameters for Scikit-learn MLP

Let's look at the key parameters for tuning the Scikit-learn MLP model. Listing 6-3 provides the example code for implementing MLPClassifier.

> ***hidden_layer_sizes***: You have to provide a number of hidden layers and neurons for each hidden layer. For example, hidden_layer_sizes – (5,3,3) means there are three hidden layers and the number of neurons for layer one is 5, for layer two it's 3, and for layer three it's 3, respectively. The default value is (100,), that is, one hidden layer with 100 neurons.

activation: This is the activation function for a hidden layer; there are four activation functions available for use; default is "relu."

- *relu*: The rectified linear unit function, returns $f(x) = \max(0, x)$

- *logistic*: The logistic sigmoid function, returns $f(x) = 1 / (1 + \exp(-x))$.

- *identity*: No-op activation, useful to implement linear bottleneck, returns $f(x) = x$

- *tanh*: The hyperbolic tan function, returns $f(x) = \tanh(x)$.

solver: This is for weight optimization. There are three options available, default being "adam."

- *adam*: Stochastic gradient-based optimizer proposed by Diederik Kingma and Jimmy Ba, which works well for a large dataset

- *lbfgs*: Belongs to the family of quasi-Newton methods, works well for small datasets

- *sgd*: Stochastic gradient descent

max_iter: This is the maximum number of iterations for the solver to converge, default is 200

learning_rate_init: This is the initial learning rate to control step-size for updating the weights (only applicable for solvers sgd/adam), default is 0.001

It is recommended to scale or normalize your data before modeling as MLP is sensitive to feature scaling.

Listing 6-3. Example Code for Sklearn MLPClassifier

```
# split data to training and testing data
X_train, X_test, y_train, y_test = train_test_split(digits.data, digits.
target, test_size=0.2, random_state=2017)
print ('Number of samples in training set: %d' %(len(y_train)))
print ('Number of samples in test set: %d' %(len(y_test)))
```

```python
# Standardise data, and fit only to the training data
scaler = StandardScaler()
scaler.fit(X_train)

# Apply the transformations to the data
X_train_scaled = scaler.transform(X_train)
X_test_scaled = scaler.transform(X_test)

# Initialize ANN classifier
mlp = MLPClassifier(hidden_layer_sizes=(100), activation='logistic',
max_iter = 100)

# Train the classifier with the training data
mlp.fit(X_train_scaled,y_train)
#----output----
Number of samples in training set: 1437
Number of samples in test set: 360

MLPClassifier(activation='logistic', alpha=0.0001, batch_size='auto',
        beta_1=0.9, beta_2=0.999, early_stopping=False, epsilon=1e-08,
        hidden_layer_sizes=(30, 30, 30), learning_rate='constant',
        learning_rate_init=0.001, max_iter=100, momentum=0.9,
        nesterovs_momentum=True, power_t=0.5, random_state=None,
        shuffle=True, solver='adam', tol=0.0001, validation_fraction=0.1,
        verbose=False, warm_start=False)

print("Training set score: %f" % mlp.score(X_train_scaled, y_train))
print("Test set score: %f" % mlp.score(X_test_scaled, y_test))
#----output----
Training set score: 0.990953
Test set score: 0.983333

# predict results from the test data
X_test_predicted = mlp.predict(X_test_scaled)

fig = plt.figure(figsize=(8, 8))  # figure size in inches
fig.subplots_adjust(left=0, right=1, bottom=0, top=1, hspace=0.05,
wspace=0.05)
```

```
# plot the digits: each image is 8x8 pixels
for i in range(32):
    ax = fig.add_subplot(8, 8, i + 1, xticks=[], yticks=[])
    ax.imshow(X_test.reshape(-1, 8, 8)[i], cmap=plt.cm.gray_r)

    # label the image with the target value
    if X_test_predicted[i] == y_test[i]:
        ax.text(0, 1, X_test_predicted[i], color='green',
        bbox=dict(facecolor='white'))
    else:
        ax.text(0, 1, X_test_predicted[i], color='red',
bbox=dict(facecolor='white'))
#----output----
```

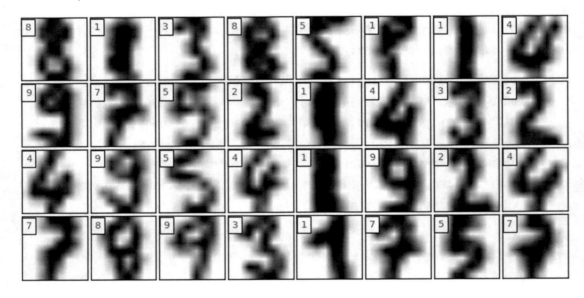

Restricted Boltzman Machines (RBMs)

An RBM algorithm was proposed by Geoffrey Hinton (2007) that learns a probability distribution over its sample training data inputs. It has seen wide applications in different areas of supervised/unsupervised ML such as feature learning, dimensionality reduction, classification, collaborative filtering, and topic modeling.

Consider the example movie rating discussed in the "Recommender Systems" section of Chapter 5. Movies like Avengers, Avatar, and Interstellar have strong associations with the latest fantasy and science fiction factor. Based on the user rating, RBM will discover latent factors that can explain the activation of movie choices. In short, RBM describes variability among correlated variables of the input dataset in terms of a potentially lower number of unobserved variables.

The energy function is given by $E(v, h) = - a^T v - b^T h - v^T W h$.

The probability function of a visible input layer can be given by

$$f(v) = -a^T v - \sum_i \log \sum_{h_i} e^{h_i(b_i + W_i v)}.$$

Let's build a logistic regression model on a digits dataset with BernoulliRBM and compare its accuracy with a straight logistic regression (without BernoulliRBM) model's accuracy.

Let's nudge the dataset by moving the 8x8 images by 1 pixel left, right, down and up to convolute the image (Listing 6-4).

Listing 6-4. Function to Nudge the Dataset

```
# Function to nudge the dataset
def nudge_dataset(X, Y):
    """

    This produces a dataset 5 times bigger than the original one,
    by moving the 8x8 images in X around by 1px to left, right, down, up
    """

    direction_vectors = [
        [[0, 1, 0],
         [0, 0, 0],
         [0, 0, 0]],

        [[0, 0, 0],
         [1, 0, 0],
         [0, 0, 0]],

        [[0, 0, 0],
         [0, 0, 1],
         [0, 0, 0]],
```

```
     [[0, 0, 0],
      [0, 0, 0],
      [0, 1, 0]]]

  shift = lambda x, w: convolve(x.reshape((8, 8)), mode='constant',
                                weights=w).ravel()
  X = np.concatenate([X] +
                     [np.apply_along_axis(shift, 1, X, vector)
                      for vector in direction_vectors])
  Y = np.concatenate([Y for _ in range(5)], axis=0)
  return X, Y
```

The BernoulliRBM assumes that the columns of our feature vectors fall within the range 0 to 1. However, the MNIST dataset is represented as unsigned 8-bit integers, falling within the range of 0 to 255.

Define a function to scale the columns into the range (0, 1). The scale function takes two parameters: our data matrix X and an epsilon value used to prevent division by zero errors (Listing 6-5).

Listing 6-5. Example Code for Using BernoulliRBM with Classifier

```
# Example adapted from scikit-learn documentation
import numpy as np
import matplotlib.pyplot as plt
from sklearn import linear_model, datasets, metrics
from sklearn.model_selection import train_test_split, GridSearchCV
from sklearn.neural_network import BernoulliRBM
from sklearn.pipeline import Pipeline
from scipy.ndimage import convolve

# Load Data
digits = datasets.load_digits()
X = np.asarray(digits.data, 'float32')
y = digits.target

X, y = nudge_dataset(X, digits.target)

# Scale the features such that the values are between 0-1 scale
X = (X - np.min(X, 0)) / (np.max(X, 0) + 0.0001)
```

```
X_train, X_test, y_train, y_test = train_test_split(X, y, test_size=0.2,
random_state=2017)
print (X.shape)
print (y.shape)
#----output----
(8985L, 64)
(8985L,)

# Gridsearch for logistic regression
# perform a grid search on the 'C' parameter of Logistic
params = {"C": [1.0, 10.0, 100.0]}

Grid_Search = GridSearchCV(LogisticRegression(), params, n_jobs = -1,
verbose = 1)
Grid_Search.fit(X_train, y_train)

# print diagnostic information to the user and grab the
print ("Best Score: %0.3f" % (Grid_Search.best_score_))

# best model
bestParams = Grid_Search.best_estimator_.get_params()

print (bestParams.items())
#----output----
Fitting 3 folds for each of 3 candidates, totalling 9 fits
Best Score: 0.774
[('warm_start', False), ('C', 100.0), ('n_jobs', 1), ('verbose', 0),
('intercept_scaling', 1), ('fit_intercept', True), ('max_iter', 100),
('penalty', 'l2'), ('multi_class', 'ovr'), ('random_state', None), ('dual',
False), ('tol', 0.0001), ('solver', 'liblinear'), ('class_weight', None)]

# evaluate using Logistic Regression and only the raw pixel
logistic = LogisticRegression(C = 100)
logistic.fit(X_train, y_train)

print ("Train accuracy: ", metrics.accuracy_score(y_train, logistic.
predict(X_train)))
print ("Test accuracyL ", metrics.accuracy_score(y_test, logistic.
predict(X_test)))
```

```
#----output----
Train accuracy:  0.797440178075
Test accuracyL  0.800779076238
```

Let's perform a grid search for an RBM + logistic regression model—a grid search on the learning rate, number of iterations, and number of components on the RBM and C for logistic regression.

Listing 6-6. Example Code for Grid Search with RBM + Logistic Regression

```
# initialize the RBM + Logistic Regression pipeline
rbm = BernoulliRBM()
logistic = LogisticRegression()
classifier = Pipeline([("rbm", rbm), ("logistic", logistic)])

params = {
    "rbm__learning_rate": [0.1, 0.01, 0.001],
    "rbm__n_iter": [20, 40, 80],
    "rbm__n_components": [50, 100, 200],
    "logistic__C": [1.0, 10.0, 100.0]}

# perform a grid search over the parameter
Grid_Search = GridSearchCV(classifier, params, n_jobs = -1, verbose = 1)
Grid_Search.fit(X_train, y_train)

# print diagnostic information to the user and grab the
# best model
print ("Best Score: %0.3f" % (gs.best_score_))

print ("RBM + Logistic Regression parameters")
bestParams = gs.best_estimator_.get_params()

# loop over the parameters and print each of them out
# so they can be manually set
for p in sorted(params.keys()):
    print ("\t %s: %f" % (p, bestParams[p]))
#----output----
Fitting 3 folds for each of 81 candidates, totalling 243 fits
Best Score: 0.505
```

```
RBM + Logistic Regression parameters
    logistic__C: 100.000000
    rbm__learning_rate: 0.001000
    rbm__n_components: 200.000000
    rbm__n_iter: 20.000000

# initialize the RBM + Logistic Regression classifier with
# the cross-validated parameters
rbm = BernoulliRBM(n_components = 200, n_iter = 20, learning_rate =
0.1,  verbose = False)
logistic = LogisticRegression(C = 100)

# train the classifier and show an evaluation report
classifier = Pipeline([("rbm", rbm), ("logistic", logistic)])
classifier.fit(X_train, y_train)

print (metrics.accuracy_score(y_train, classifier.predict(X_train)))
print (metrics.accuracy_score(y_test, classifier.predict(X_test)))
#----output----
0.936839176405
0.932109070673

# plot RBM components
plt.figure(figsize=(15, 15))
for i, comp in enumerate(rbm.components_):
    plt.subplot(20, 20, i + 1)
    plt.imshow(comp.reshape((8, 8)), cmap=plt.cm.gray_r,
               interpolation='nearest')
    plt.xticks(())
    plt.yticks(())
plt.suptitle('200 components extracted by RBM', fontsize=16)
plt.show()
#----output----
```

200 components extracted by RBM

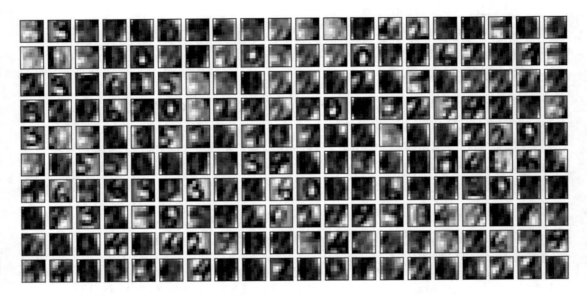

Notice that the logistic regression model with RBM lifts the model score by more than 10% compared with the model without RBM.

Note   To practice further and get a better understanding, I recommend you to try the preceding example code on Scikit-learn's Olivetti faces dataset, which contains face images taken between April 1992 and April 1994 at AT&T Laboratories Cambridge. You can load the data using `olivetti = datasets.fetch_olivetti_faces()`.

Stacked RBM is known as a deep believe network (DBN), which is an initialization technique. However, this technique was popular during 2006-2007 but is reasonably outdated. So there is no out-of-the-box implementation of DBN in Keras. However, if you are interested in a simple DBN implementation, I recommend having a look at `https://github.com/albertbup/deep-belief-network`, which has MIT license.

MLP Using Keras

In Keras, neural networks are defined as a sequence of layers, and the container for these layers is the sequential class. The sequential model is a linear stack of layers; each layer's output feeds into the next layer's input.

The first layer in the neural network will define the number of inputs to expect. The activation functions transform a summed signal from each neuron in a layer; the same can be extracted and added to the sequential as a layer-like object called activation. The choice of action depends on the type of problem (like regression or binary classification or multiclass classification) that we are trying to address (Listing 6-7).

Listing 6-7. Example Code for Keras MLP

```
from matplotlib import pyplot as plt
import numpy as np
np.random.seed(2017)

from keras.models import Sequential
from keras.datasets import mnist
from keras.layers import Dense, Activation, Dropout, Input
from keras.models import Model
from keras.utils import np_utils

from IPython.display import SVG
from keras import backend as K
from keras.callbacks import EarlyStopping
from keras.utils.visualize_util import model_to_dot, plot_model

# load data
(X_train, y_train), (X_test, y_test) = mnist.load_data()

X_train = X_train.reshape(X_train.shape[0], input_unit_size)
X_test  = X_test.reshape(X_test.shape[0], input_unit_size)
X_train = X_train.astype('float32')
X_test  = X_test.astype('float32')

# Scale the values by dividing 255 i.e., means foreground (black)
X_train /= 255
X_test  /= 255

# one-hot representation, required for multiclass problems
y_train = np_utils.to_categorical(y_train, nb_classes)
y_test = np_utils.to_categorical(y_test, nb_classes)
```

```
print('X_train shape:', X_train.shape)
print(X_train.shape[0], 'train samples')
print(X_test.shape[0], 'test samples')

nb_classes = 10 # class size
# flatten 28*28 images to a 784 vector for each image
input_unit_size = 28*28

# create model
model = Sequential()
model.add(Dense(input_unit_size, input_dim=input_unit_size, kernel_
initializer='normal', activation='relu'))
model.add(Dense(nb_classes, kernel_initializer='normal',
activation='softmax'))
#----output----
'X_train shape:', (60000, 784)
60000, 'train samples'
10000, 'test samples'
```

The compilation is a model with a precompute step that transforms the sequence of layers that we defined into a highly efficient series of matrix transforms. It takes three arguments: an optimizer, a loss function, and a list of evaluation metrics.

Unlike Scikit-learn implementation, Keras provides a rich number of optimizers such as SGD, RMSprop, Adagrad (adaptive subgradient), Adadelta (adaptive learning rate), Adam, Adamax, Nadam, and TFOptimizer. For brevity, I'll not explain these here but refer you to the official Keras site for further reference.

Some standard loss functions are "mse" for regression, binary_crossentropy (logarithmic loss) for binary classification, and categorical_crossentropy (multiclass logarithmic loss) for multiclassification problems.

The standard evaluation metrics for different types of problems are supported, and you can pass a list to them to evaluate (Listing 6-8).

Listing 6-8. Compile the Model

```
# Compile model
model.compile(loss='categorical_crossentropy', optimizer='adam',
metrics=['accuracy'])

SVG(model_to_dot(model, show_shapes=True).create(prog='dot', format='svg'))
```

The network is trained using backpropagation algorithm and optimized according to the specified method and loss function. Each epoch can be partitioned into batches.

Listing 6-9. Train Model and Evaluate

```
# model training
model.fit(X_train, y_train, validation_data=(X_test, y_test), nb_epoch=5,
batch_size=500, verbose=2)

# Final evaluation of the model
scores = model.evaluate(X_test, y_test, verbose=0)
print("Error: %.2f%%" % (100-scores[1]*100))
#----output----
Train on 60000 samples, validate on 10000 samples
Epoch 1/5
 - 3s - loss: 0.3863 - acc: 0.8926 - val_loss: 0.1873 - val_acc: 0.9477
Epoch 2/5
 - 3s - loss: 0.1558 - acc: 0.9561 - val_loss: 0.1280 - val_acc: 0.9612
Epoch 3/5
 - 3s - loss: 0.1071 - acc: 0.9696 - val_loss: 0.1009 - val_acc: 0.9697
Epoch 4/5
 - 3s - loss: 0.0800 - acc: 0.9773 - val_loss: 0.0845 - val_acc: 0.9756
Epoch 5/5
 - 3s - loss: 0.0607 - acc: 0.9832 - val_loss: 0.0760 - val_acc: 0.9776
Error: 2.24%
```

Listing 6-10. Additional Example to Train Model and Evaluate for Diabetes Dataset

```
import pandas as pd

# load pima indians dataset
dataset = pd.read_csv('Data/Diabetes.csv')

# split into input (X) and output (y) variables
X = dataset.iloc[:,0:8].values
y = dataset['class'].values      # dependent variables
```

```
# create model
model = Sequential()
model.add(Dense(12, input_dim=8, kernel_initializer='uniform',
activation='relu'))
model.add(Dense(1, kernel_initializer='uniform', activation='sigmoid'))

# Compile model
model.compile(loss='binary_crossentropy', optimizer='adam',
metrics=['accuracy'])

SVG(model_to_dot(model, show_shapes=True).create(prog='dot', format='svg'))

# Fit the model
model.fit(X, y, epochs=5, batch_size=10)
# evaluate the model
scores = model.evaluate(X, y)
print("%s: %.2f%%" % (model.metrics_names[1], scores[1]*100))
#----output----
Epoch 1/5
768/768 [==============================] - 0s 306us/step - loss:
0.6737 - acc: 0.6250
Epoch 2/5
768/768 [==============================] - 0s 118us/step - loss:
0.6527 - acc: 0.6510
Epoch 3/5
768/768 [==============================] - 0s 96us/step - loss:
0.6432 - acc: 0.6563
Epoch 4/5
768/768 [==============================] - 0s 109us/step - loss:
0.6255 - acc: 0.6719
Epoch 5/5
768/768 [==============================] - 0s 113us/step - loss:
0.6221 - acc: 0.6706
768/768 [==============================] - 0s 84us/step
acc: 68.75%
```

Autoencoders

As the name suggests, an autoencoder aims to learn to encode a representation of a training sample data automatically without human intervention. The autoencoder is widely used for dimensionality reduction and data denoising (Figure 6-5).

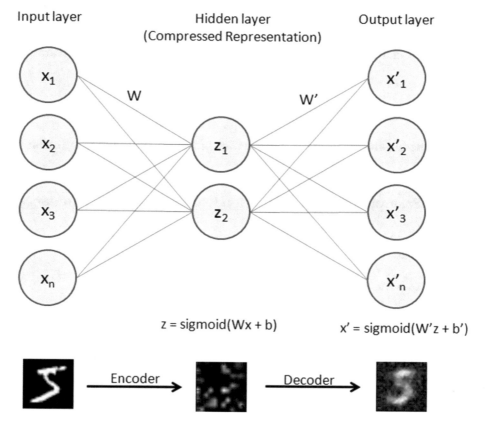

Figure 6-5. *Autoencoder*

Building an autoencoder will typically have three elements:

1. Encoding function to map input to a hidden representation through a nonlinear function, $z = $ sigmoid $(Wx + b)$

2. A decoding function such as $x' = $ sigmoid$(W'y + b')$, which will map back into reconstruction x' with the same shape as x

3. A loss function, which is a distance function to measure the
    information loss between the compressed representation of data
    and the decompressed representation. Reconstruction error can
    be measured using traditional squared error $||x-z||^2$.

We'll be using the well-known MNIST database of handwritten digits, which
consists of approximately 70,000 total samples of handwritten grayscale digit images for
numbers 0 to 9. Each image is of 28×28 size and intensity level varies from 0 to 255, with
accompanying label integer 0 to 9 for 60,000 of them and the remainder without label
(test data set).

Dimension Reduction Using an Autoencoder

Listing 6-11 provides an example code implementation for reducing dimension using an
autoencoder.

Listing 6-11. Example Code for Dimension Reduction Using an Autoencoder

```
import numpy as np
np.random.seed(2017)

from keras.datasets import mnist
from keras.models import Model
from keras.layers import Input, Dense
from keras.optimizers import Adadelta
from keras.utils import np_utils

from IPython.display import SVG
from keras import backend as K
from keras.callbacks import EarlyStopping
from keras.utils.visualize_util import model_to_dot
from matplotlib import pyplot as plt

# Load mnist data
input_unit_size = 28*28
(X_train, y_train), (X_test, y_test) = mnist.load_data()
```

```python
# function to plot digits
def draw_digit(data, row, col, n):
    size = int(np.sqrt(data.shape[0]))
    plt.subplot(row, col, n)
    plt.imshow(data.reshape(size, size))
    plt.gray()

# Normalize
X_train = X_train.reshape(X_train.shape[0], input_unit_size)
X_train = X_train.astype('float32')
X_train /= 255
print('X_train shape:', X_train.shape)
#----output----
'X_train shape:', (60000, 784)

# Autoencoder
inputs = Input(shape=(input_unit_size,))
x = Dense(144, activation='relu')(inputs)
outputs = Dense(input_unit_size)(x)
model = Model(input=inputs, output=outputs)
model.compile(loss='mse', optimizer='adadelta')

SVG(model_to_dot(model, show_shapes=True).create(prog='dot', format='svg'))
#----output----
```

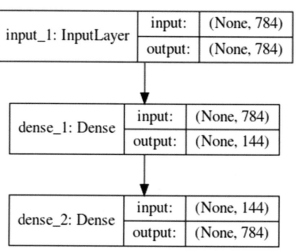

Note that the 784 dimension is reduced through encoding to 144 in the hidden layer and again in layer 3 constructed back to 784 using the decoder.

```
model.fit(X_train, X_train, nb_epoch=5, batch_size=258)
#----output----
Epoch 1/5
60000/60000 [==============================] - 8s - loss: 0.0733
Epoch 2/5
60000/60000 [==============================] - 9s - loss: 0.0547
Epoch 3/5
60000/60000 [==============================] - 11s - loss: 0.0451
Epoch 4/5
60000/60000 [==============================] - 11s - loss: 0.0392
Epoch 5/5
60000/60000 [==============================] - 11s - loss: 0.0354

# plot the images from input layers
show_size = 5
total = 0
plt.figure(figsize=(5,5))
for i in range(show_size):
    for j in range(show_size):
        draw_digit(X_train[total], show_size, show_size, total+1)
        total+=1
plt.show()
#----output----
```

```
# plot the encoded (compressed) layer image
get_layer_output = K.function([model.layers[0].input],
                              [model.layers[1].output])

hidden_outputs = get_layer_output([X_train[0:show_size**2]])[0]

total = 0
plt.figure(figsize=(5,5))
for i in range(show_size):
    for j in range(show_size):
        draw_digit(hidden_outputs[total], show_size, show_size, total+1)
        total+=1
plt.show()
#----output----
```

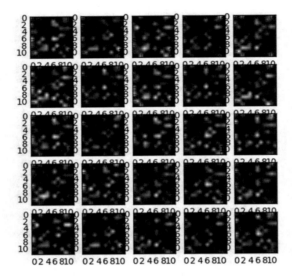

```
# Plot the decoded (de-compressed) layer images
get_layer_output = K.function([model.layers[0].input],
                              [model.layers[2].output])

last_outputs = get_layer_output([X_train[0:show_size**2]])[0]

total = 0
plt.figure(figsize=(5,5))
for i in range(show_size):
    for j in range(show_size):
        draw_digit(last_outputs[total], show_size, show_size, total+1)
        total+=1
plt.show()
#----output----
```

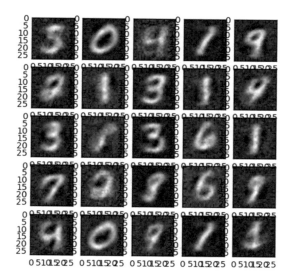

Denoise Image Using an Autoencoder

Discovering robust features from the compressed hidden layer is an important aspect
to enable the autoencoder to efficiently reconstruct the input from a denoised version
or original image. This is addressed by the denoising autoencoder, which is a stochastic
version of an autoencoder.

Let's introduce some noise to the digit dataset and try to build a model to denoise
the image (Listing 6-12).

Listing 6-12. Example Code for Denoising Using an Autoencoder

```
# Introducing noise to the image
noise_factor = 0.5
X_train_noisy = X_train + noise_factor * np.random.normal(loc=0.0,
scale=1.0, size=X_train.shape)
X_train_noisy = np.clip(X_train_noisy, 0., 1.)

# Function for visualization
def draw(data, row, col, n):
    plt.subplot(row, col, n)
    plt.imshow(data, cmap=plt.cm.gray_r)
    plt.axis('off')
```

```
show_size = 10
plt.figure(figsize=(20,20))

for i in range(show_size):
    draw(X_train_noisy[i].reshape(28,28), 1, show_size, i+1)
plt.show()
#----output----
```

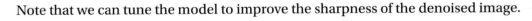

```
#Let's fit a model on noisy training dataset.
model.fit(X_train_noisy, X_train, nb_epoch=5, batch_size=258)

# Prediction for denoised image
X_train_pred = model.predict(X_train_noisy)

show_size = 10
plt.figure(figsize=(20,20))

for i in range(show_size):
    draw(X_train_pred[i].reshape(28,28), 1, show_size, i+1)
plt.show()
#----output----
```

Note that we can tune the model to improve the sharpness of the denoised image.

Convolutional Neural Network (CNN)

In the world of image classification, the CNN has become the go-to algorithm to build efficient models. CNNs are similar to an ordinary neural network, except that it explicitly assumes that the inputs are images, which allows us to encode certain properties into the architecture. These then make the forward function efficient, to implement and reduces the parameters in the network. The neurons are arranged in three dimensions: width, height, and depth.

Let's consider CIFAR-10 (Canadian Institute for Advanced Research), which is a standard computer vision and deep learning image dataset. It consists of 60,000 color photos of 32 by 32 pixels squared, with RGB for each pixel, divided into ten classes that include common objects such as airplanes, automobiles, birds, cats, deer, dog, frog, horse, ship, and truck. Essentially each image is of size 32×32×3 (width × height × RGB color channels).

A CNN consists of four main types of layers: input layer, convolution layer, pooling layer, and fully connected layer.

The input layer will hold the raw pixel, so an image of CIFAR-10 will have 32×32×3 dimensions at the input layer. The convolution layer will compute a dot product between the weights of a small local region from the input layer, so if we decide to have five filters, the resultant reduced dimension will be 32×32×5. The ReLU layer will apply an elementwise activation function, which will not affect the dimension. The pool layer will down sample the spatial dimension along width and height, resulting in dimensions of 16×16×5. Finally, the fully connected layer will compute the class score, and the resultant dimension will be a single vector 1×1×10 (ten class scores). Each neuron in this layer is connected to all numbers in the previous volume (Figure 6-6).

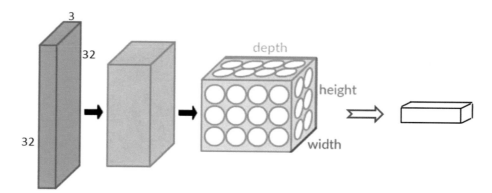

Figure 6-6. *Convolutional neural network*

The following example illustration uses Keras with Theano backend. To start Kearas with Theano backend, please run the following command while starting Jupyter Notebook, "KERAS_BACKEND=theano jupyter notebook" (Listing 6-13).

Listing 6-13. CNN Using Keras with Theano Backend on CIFAR10 Dataset

```python
import keras
if K=='tensorflow':
    keras.backend.set_image_dim_ordering('tf')
else:
    keras.backend.set_image_dim_ordering('th')

from keras.models import Sequential
from keras.datasets import cifar10
from keras.layers import Dense, Dropout, Activation, Conv2D, MaxPooling2D,
Flatten
from keras.utils import np_utils
from keras.preprocessing import sequence

from keras import backend as K
from IPython.display import SVG, display
from keras.utils.vis_utils import model_to_dot, plot_model
import numpy as np
np.random.seed(2017)

img_rows, img_cols = 32, 32
img_channels = 3

batch_size = 256
nb_classes = 10
nb_epoch = 4
nb_filters = 10
nb_conv = 3
nb_pool = 2
kernel_size = 3 # convolution kernel size

if K.image_dim_ordering() == 'th':
    input_shape = (3, img_rows, img_cols)
else:
    input_shape = (img_rows, img_cols, 3)
```

```
(X_train, y_train), (X_test, y_test) = cifar10.load_data()
print('X_train shape:', X_train.shape)
print(X_train.shape[0], 'train samples')
print(X_test.shape[0], 'test samples')
X_train = X_train.astype('float32')
X_test = X_test.astype('float32')
X_train /= 255
X_test /= 255

Y_train = np_utils.to_categorical(y_train, nb_classes)
Y_test = np_utils.to_categorical(y_test, nb_classes)
#----output----
X_train shape: (50000, 32, 32, 3)
50000 train samples
10000 test samples

# define two groups of layers: feature (convolutions) and classification (dense)
feature_layers = [
    Conv2D(nb_filters, kernel_size, input_shape=input_shape),
    Activation('relu'),
    Conv2D(nb_filters, kernel_size),
    Activation('relu'),
    MaxPooling2D(pool_size=(nb_pool, nb_pool)),
    Flatten(),
]
classification_layers = [
    Dense(512),
    Activation('relu'),
    Dense(nb_classes),
    Activation('softmax')
]

# create complete model
model = Sequential(feature_layers + classification_layers)

model.compile(loss='categorical_crossentropy', optimizer="adadelta",
metrics=['accuracy'])
```

```
SVG(model_to_dot(model, show_shapes=True).create(prog='dot', format='svg'))
#----output----
```

Layer (type)	Output Shape	Param #
conv2d_1 (Conv2D)	(None, 30, 30, 10)	280
activation_1 (Activation)	(None, 30, 30, 10)	0
conv2d_2 (Conv2D)	(None, 28, 28, 10)	910
activation_2 (Activation)	(None, 28, 28, 10)	0
max_pooling2d_1 (MaxPooling2	(None, 14, 14, 10)	0
flatten_1 (Flatten)	(None, 1960)	0
dense_1 (Dense)	(None, 512)	1004032
activation_3 (Activation)	(None, 512)	0
dense_2 (Dense)	(None, 10)	5130
activation_4 (Activation)	(None, 10)	0

```
Total params: 1,010,352
Trainable params: 1,010,352
Non-trainable params: 0

# fit model
model.fit(X_train, Y_train, validation_data=(X_test, Y_test),
          epochs=nb_epoch, batch_size=batch_size, verbose=2)
#----output----
Train on 50000 samples, validate on 10000 samples
Epoch 1/4
 - 50s - loss: 1.8512 - acc: 0.3422 - val_loss: 1.5729 - val_acc: 0.4438
```

```
Epoch 2/4
 - 38s - loss: 1.4350 - acc: 0.4945 - val_loss: 1.4312 - val_acc: 0.4832
Epoch 3/4
 - 38s - loss: 1.2542 - acc: 0.5566 - val_loss: 1.3300 - val_acc: 0.5191
Epoch 4/4
 - 38s - loss: 1.1375 - acc: 0.6021 - val_loss: 1.1760 - val_acc: 0.5760
```

Let's visualize each layer. Note that we applied ten filters.

```
# function for Visualization
# visualization
def draw(data, row, col, n):
    plt.subplot(row, col, n)
    plt.imshow(data)

def draw_digit(data, row, col):
    for j in range(row):
        plt.figure(figsize=(16,16))
        for i in range(col):
            plt.subplot(row, col, i+1)
            plt.imshow(data[j,:,:,i])
            plt.axis('off')
        plt.tight_layout()
    plt.show()

### Input layer (original image)
show_size = 10
plt.figure(figsize=(16,16))
for i in range(show_size):
    draw(X_train[i], 1, show_size, i+1)
plt.show()
#----output----
```

Notice in the following that the hidden layers features are stored in ten filters.

```
# first layer
get_first_layer_output = K.function([model.layers[0].input],
                          [model.layers[1].output])
first_layer = get_first_layer_output([X_train[0:show_size]])[0]

print ('first layer shape: ', first_layer.shape)
draw_digit(first_layer, first_layer.shape[0], first_layer.shape[3])
#----output----
```

```
# second layer
get_second_layer_output = K.function([model.layers[0].input],
                            [model.layers[3].output])
second_layers = get_second_layer_output([X_train[0:show_size]])[0]

print ('second layer shape: ', second_layers.shape)
draw_digit(second_layers, second_layers.shape[0], second_layers.shape[3])
#----output----
```

```
# third layer
get_third_layer_output = K.function([model.layers[0].input],
                           [model.layers[4].output])
third_layers = get_third_layer_output([X_train[0:show_size]])[0]

print ('third layer shape: ', third_layers.shape)
draw_digit(third_layers, third_layers.shape[0], third_layers.shape[3])
#----output-----
```

CNN on MNIST Dataset

As an additional example, let's look at how the CNN might look on the digits dataset (Listing 6-14).

Listing 6-14. CNN Using Keras with Theano Backend on MNIST Dataset

```
import keras
keras.backend.backend()
keras.backend.image_dim_ordering()

# using theano as backend
K = keras.backend.backend()
if K=='tensorflow':
    keras.backend.set_image_dim_ordering('tf')
else:
    keras.backend.set_image_dim_ordering('th')

from matplotlib import pyplot as plt
%matplotlib inline

import numpy as np
np.random.seed(2017)

from keras import backend as K
from keras.models import Sequential
from keras.datasets import mnist
from keras.layers import Dense, Dropout, Activation, Conv2D, MaxPooling2D,
Flatten
from keras.utils import np_utils
from keras.preprocessing import sequence

from keras import backend as K
from IPython.display import SVG, display
from keras.utils.vis_utils import model_to_dot, plot_model

nb_filters = 5 # the number of filters
nb_pool = 2 # window size of pooling
nb_conv = 3 # window or kernel size of filter
```

```
nb_epoch = 5
kernel_size = 3 # convolution kernel size

if K.image_dim_ordering() == 'th':
    input_shape = (1, img_rows, img_cols)
else:
    input_shape = (img_rows, img_cols, 1)

# data
(X_train, y_train), (X_test, y_test) = mnist.load_data()

X_train = X_train.reshape(X_train.shape[0], img_rows, img_cols, 1)
X_test = X_test.reshape(X_test.shape[0], img_rows, img_cols, 1)
X_train = X_train.astype('float32')
X_test = X_test.astype('float32')
X_train /= 255
X_test /= 255
print('X_train shape:', X_train.shape)
print(X_train.shape[0], 'train samples')
print(X_test.shape[0], 'test samples')

# convert class vectors to binary class matrices
Y_train = np_utils.to_categorical(y_train, nb_classes)
Y_test = np_utils.to_categorical(y_test, nb_classes)
#----output----
'X_train shape:', (60000, 1, 28, 28)
60000, 'train samples'
10000, 'test samples'

# define two groups of layers: feature (convolutions) and classification
(dense)
feature_layers = [
    Conv2D(nb_filters, kernel_size, input_shape=input_shape),
    Activation('relu'),
    Conv2D(nb_filters, kernel_size),
    Activation('relu'),
```

```
    MaxPooling2D(pool_size = nb_pool),
    Dropout(0.25),
    Flatten(),
]
classification_layers = [
    Dense(128),
    Activation('relu'),
    Dropout(0.5),
    Dense(nb_classes),
    Activation('softmax')
]

# create complete model
model = Sequential(feature_layers + classification_layers)

model.compile(loss='categorical_crossentropy', optimizer="adadelta",
metrics=['accuracy'])

SVG(model_to_dot(model, show_shapes=True).create(prog='dot', format='svg'))

print(model.summary())
#----output----
```

Layer (type)	Output Shape	Param #
conv2d_1 (Conv2D)	(None, 26, 26, 5)	50
activation_1 (Activation)	(None, 26, 26, 5)	0
conv2d_2 (Conv2D)	(None, 24, 24, 5)	230
activation_2 (Activation)	(None, 24, 24, 5)	0
max_pooling2d_1 (MaxPooling2	(None, 12, 12, 5)	0
dropout_1 (Dropout)	(None, 12, 12, 5)	0

flatten_1 (Flatten)	(None, 720)	0
dense_1 (Dense)	(None, 128)	92288
activation_3 (Activation)	(None, 128)	0
dropout_2 (Dropout)	(None, 128)	0
dense_2 (Dense)	(None, 10)	1290
activation_4 (Activation)	(None, 10)	0

```
=================================================================
Total params: 93,858
Trainable params: 93,858
Non-trainable params: 0

model.fit(X_train, Y_train, batch_size=256, epochs=nb_epoch,
verbose=2,  validation_split=0.2)
#----output----
Train on 48000 samples, validate on 12000 samples
Epoch 1/5
 - 15s - loss: 0.6098 - acc: 0.8085 - val_loss: 0.1609 - val_acc: 0.9523
Epoch 2/5
 - 15s - loss: 0.2427 - acc: 0.9251 - val_loss: 0.1148 - val_acc: 0.9675
Epoch 3/5
 - 15s - loss: 0.1941 - acc: 0.9410 - val_loss: 0.0950 - val_acc: 0.9727
Epoch 4/5
 - 15s - loss: 0.1670 - acc: 0.9483 - val_loss: 0.0866 - val_acc: 0.9753
Epoch 5/5
 - 15s - loss: 0.1500 - acc: 0.9548 - val_loss: 0.0830 - val_acc: 0.9767
```

Visualization of Layers

```
# visualization
def draw(data, row, col, n):
    plt.subplot(row, col, n)
```

```
        plt.imshow(data, cmap=plt.cm.gray_r)
        plt.axis('off')

 def draw_digit(data, row, col):
        for j in range(row):
            plt.figure(figsize=(8,8))
            for i in range(col):
                plt.subplot(row, col, i+1)
                plt.imshow(data[j,:,:,i], cmap=plt.cm.gray_r)
                plt.axis('off')
            plt.tight_layout()
        plt.show()

# Sample input layer (original image)
show_size = 10
plt.figure(figsize=(20,20))

for i in range(show_size):
    draw(X_train[i].reshape(28,28), 1, show_size, i+1)
plt.show()
#----output----
```

```
# First layer with 5 filters
get_first_layer_output = K.function([model.layers[0].input], [model.
layers[1].output])
first_layer = get_first_layer_output([X_train[0:show_size]])[0]

print ('first layer shape: ', first_layer.shape)

draw_digit(first_layer, first_layer.shape[0], first_layer.shape[3])
#----output----
```

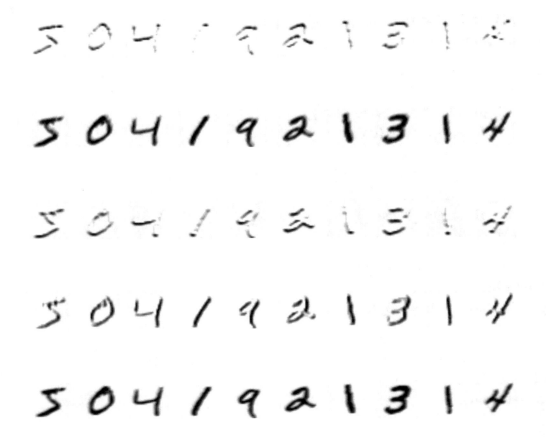

Recurrent Neural Network (RNN)

The MLP (feedforward network) is not known to do well on a sequential events model—such as the probabilistic language model—for predicting the next word based on the previous word at every given point. RNN architecture addresses this issue. It is similar to MLP except that it has a feedback loop, which means it feeds previous time steps into the current step. This type of architecture generates sequences to simulate the situation and create synthetic data. This makes it the ideal modeling choice to work on sequence data such as speech text mining, image captioning, time series prediction, robot control, language modeling, etc. (Figure 6-7).

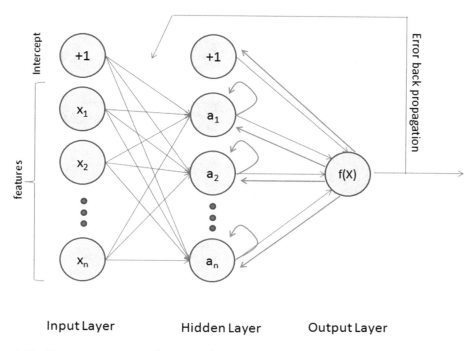

Figure 6-7. *Recurrent neural network*

The previous step's hidden layer and final outputs are fed back into the network and will be used as input to the next step's hidden layer, which means the network will remember the past and it will repeatedly predict what will happen next. The drawback in the general RNN architecture is that it can be memory heavy, and hard to train for long-term temporal dependency (i.e., the context of long text should be known at any given stage).

Long Short Term Memory (LSTM)

LSTM is an implementation of improved RNN architecture to address the issues of general RNN, and it enables long-range dependencies. It is designed to have better memory through linear memory cells surrounded by a set of gate units used to control the flow of information—when information should enter the memory, when to forget, and when to output. It uses no activation function within its recurrent components, thus the gradient term does not vanish with backpropagation. Figure 6-8 gives a comparison of a simple multilayer perceptron vs. RNN vs. LSTM.

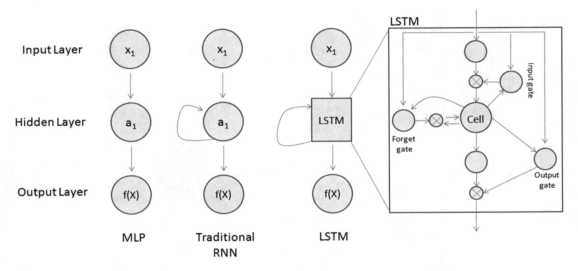

Figure 6-8. *Simple MLP vs. RNN vs. LSTM*

Please refer to Table 6-3 to understand the key LSTM component formulas.

Table 6-3. *LSTM Components*

LSTM Component	Formula
Input gate layer: This decides which values to store in the cell state.	$i_t = \text{sigmoid}(w_i x_t + u_i h_{t-1} + b_i)$
Forget gate layer: As the name suggests, this decides what information to throw away from the cell state.	$f_t = \text{sigmoid}(W_f x_t + U_f h_{t-1} + b_f)$
Output gate layer: Creates a vector of values that can be added to the cell state.	$0_t = \text{sigmoid}(W_o x_t + u_i h_{t-1} + b_o)$
Memory cell state vector	$c_t = f_t\ 0\ c_{t-1} + i_t 0 * \text{hyperbolic tangent}(W_c x_t + u_c h_{t-1} + b_c)$

Let's look at an example of an IMDB dataset that has labeled sentiment (positive/negative) for movie reviews. The reviews have been preprocessed, and encoded as a sequence of word indexes (Listing 6-15).

Listing 6-15. Example Code for Keras LSTM

```
import numpy as np
np.random.seed(2017)  # for reproducibility

from keras.preprocessing import sequence
from keras.models import Sequential
from keras.layers import Dense, Activation, Embedding
from keras.layers import LSTM
from keras.datasets import imdb

max_features = 20000
maxlen = 80  # cut texts after this number of words (among top max_features
most common words)
batch_size = 32

print('Loading data...')
(X_train, y_train), (X_test, y_test) = imdb.load_data(num_words=max_
features)
print(len(X_train), 'train sequences')
print(len(X_test), 'test sequences')

print('Pad sequences (samples x time)')
X_train = sequence.pad_sequences(X_train, maxlen=maxlen)
X_test = sequence.pad_sequences(X_test, maxlen=maxlen)
print('X_train shape:', X_train.shape)
print('X_test shape:', X_test.shape)
#----output----
Loading data...
25000 train sequences
25000 test sequences
Pad sequences (samples x time)
X_train shape: (25000, 80)
X_test shape: (25000, 80)

#Model configuration
model = Sequential()
model.add(Embedding(max_features, 128))
```

```
model.add(LSTM(128, recurrent_dropout=0.2, dropout=0.2))  # try using a GRU
instead, for fun
model.add(Dense(1))
model.add(Activation('sigmoid'))

# Try using different optimizers and different optimizer configs
model.compile(loss='binary_crossentropy', optimizer='adam',
metrics=['accuracy'])

#Train
model.fit(X_train, y_train, batch_size=batch_size, epochs=5, validation_
data=(X_test, y_test))
#----output----
Epoch 1/5
25000/25000 [==============================] - 99s 4ms/step - loss:
0.4604 - acc: 0.7821 - val_loss: 0.3762 - val_acc: 0.8380
Epoch 2/5
25000/25000 [==============================] - 86s 3ms/step - loss:
0.3006 - acc: 0.8766 - val_loss: 0.3710 - val_acc: 0.8353
Epoch 3/5
25000/25000 [==============================] - 86s 3ms/step - loss:
0.2196 - acc: 0.9146 - val_loss: 0.4113 - val_acc: 0.8212
Epoch 4/5
25000/25000 [==============================] - 86s 3ms/step - loss:
0.1558 - acc: 0.9411 - val_loss: 0.4733 - val_acc: 0.8116
Epoch 5/5
25000/25000 [==============================] - 86s 3ms/step - loss:
0.1112 - acc: 0.9597 - val_loss: 0.6225 - val_acc: 0.8202

# Evaluate
train_score, train_acc = model.evaluate(X_train, y_train, batch_
size=batch_size)
test_score, test_acc = model.evaluate(X_test, y_test, batch_size=batch_size)

print ('Train score:', train_score)
print ('Train accuracy:', train_acc)
```

```
print ('Test score:', test_score)
print ('Test accuracy:', test_acc)
#----output----
25000/25000 [==============================] - 37s 1ms/step
25000/25000 [==============================] - 28s 1ms/step
Train score: 0.055540263980031014
Train accuracy: 0.98432
Test score: 0.5643649271917344
Test accuracy: 0.82388
```

Transfer Learning

Based on our past experience, we humans can learn a new skill easily. We are more efficient in learning, particularly if the task at hand is similar to what we have done in the past. For example, learning a new programming language for a computer professional or driving a new type of vehicle for a seasoned driver is relatively easy, based on our past experience.

Transfer learning is an area in ML that aims to utilize the knowledge gained while solving one problem to solve a different but related problem (Figure 6-9).

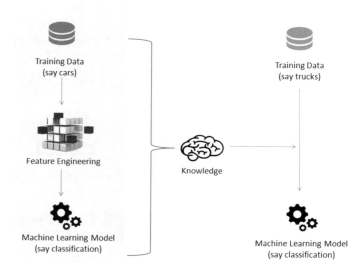

Figure 6-9. *Transfer learning*

There's nothing better than understanding through example, so let's train a simple CNN model of two level layers (a feature layer and a classification layer) on the first 5 digits (0 to 4) of the MNIST dataset, then apply transfer learning to freeze the features layer and fine-tune dense layers for the classification of digits 5 to 9 (Listing 6-16).

Listing 6-16. Example Code for Transfer Learning

```
import numpy as np
np.random.seed(2017)  # for reproducibility

from keras.datasets import mnist
from keras.models import Sequential
from keras.layers import Dense, Dropout, Activation, Flatten
from keras.layers import Conv2D, MaxPooling2D
from keras.utils import np_utils
from keras import backend as K

batch_size = 128
nb_classes = 5
nb_epoch = 5

# input image dimensions
img_rows, img_cols = 28, 28

# number of convolutional filters to use
nb_filters = 32

# size of pooling area for max pooling
pool_size = 2

# convolution kernel size
kernel_size = 3

input_shape = (img_rows, img_cols, 1)

# the data, shuffled and split between train and test sets
(X_train, y_train), (X_test, y_test) = mnist.load_data()

# create two datasets one with digits below 5 and one with 5 and above
X_train_lt5 = X_train[y_train < 5]
y_train_lt5 = y_train[y_train < 5]
```

```
X_test_lt5 = X_test[y_test < 5]
y_test_lt5 = y_test[y_test < 5]

X_train_gte5 = X_train[y_train >= 5]
y_train_gte5 = y_train[y_train >= 5] - 5   # make classes start at 0 for
X_test_gte5 = X_test[y_test >= 5]             # np_utils.to_categorical
y_test_gte5 = y_test[y_test >= 5] - 5

# Train model for digits 0 to 4
def train_model(model, train, test, nb_classes):
    X_train = train[0].reshape((train[0].shape[0],) + input_shape)
    X_test = test[0].reshape((test[0].shape[0],) + input_shape)
    X_train = X_train.astype('float32')
    X_test = X_test.astype('float32')
    X_train /= 255
    X_test /= 255
    print('X_train shape:', X_train.shape)
    print(X_train.shape[0], 'train samples')
    print(X_test.shape[0], 'test samples')

    # convert class vectors to binary class matrices
    Y_train = np_utils.to_categorical(train[1], nb_classes)
    Y_test = np_utils.to_categorical(test[1], nb_classes)

    model.compile(loss='categorical_crossentropy',
                optimizer='adadelta',
                metrics=['accuracy'])

    model.fit(X_train, Y_train,
            batch_size=batch_size, epochs=nb_epoch,
            verbose=1,
            validation_data=(X_test, Y_test))
    score = model.evaluate(X_test, Y_test, verbose=0)
    print('Test score:', score[0])
    print('Test accuracy:', score[1])
```

```
# define two groups of layers: feature (convolutions) and classification
  (dense)
feature_layers = [
    Conv2D(nb_filters, kernel_size,
                   padding='valid',
                   input_shape=input_shape),
    Activation('relu'),
    Conv2D(nb_filters, kernel_size),
    Activation('relu'),
    MaxPooling2D(pool_size=(pool_size, pool_size)),
    Dropout(0.25),
    Flatten(),
]
classification_layers = [
    Dense(128),
    Activation('relu'),
    Dropout(0.5),
    Dense(nb_classes),
    Activation('softmax')
]

# create complete model
model = Sequential(feature_layers + classification_layers)

# train model for 5-digit classification [0..4]
train_model(model, (X_train_lt5, y_train_lt5), (X_test_lt5, y_test_lt5),
nb_classes)
#----output----
X_train shape: (30596, 28, 28, 1)
30596 train samples
5139 test samples
Train on 30596 samples, validate on 5139 samples
Epoch 1/5
30596/30596 [==============================] - 40s 1ms/step - loss:
0.1692 - acc: 0.9446 - val_loss: 0.0573 - val_acc: 0.9798
Epoch 2/5
```

```
30596/30596 [==============================] - 37s 1ms/step - loss:
0.0473 - acc: 0.9858 - val_loss: 0.0149 - val_acc: 0.9947
Epoch 3/5
30596/30596 [==============================] - 37s 1ms/step - loss:
0.0316 - acc: 0.9906 - val_loss: 0.0112 - val_acc: 0.9947
Epoch 4/5
30596/30596 [==============================] - 37s 1ms/step - loss:
0.0257 - acc: 0.9928 - val_loss: 0.0094 - val_acc: 0.9967
Epoch 5/5
30596/30596 [==============================] - 37s 1ms/step - loss:
0.0204 - acc: 0.9940 - val_loss: 0.0078 - val_acc: 0.9977
Test score: 0.00782204038783338
Test accuracy: 0.9976649153531816
```

Transfer existing trained model on 0 to 4 to build model for digits 5 to 9

```
# freeze feature layers and rebuild model
for layer in feature_layers:
    layer.trainable = False
```

```
# transfer: train dense layers for new classification task [5..9]
train_model(model, (X_train_gte5, y_train_gte5), (X_test_gte5,
y_test_gte5), nb_classes)
#----output----
X_train shape: (29404, 28, 28, 1)
29404 train samples
4861 test samples
Train on 29404 samples, validate on 4861 samples
Epoch 1/5
29404/29404 [==============================] - 14s 484us/step - loss:
0.2290 - acc: 0.9353 - val_loss: 0.0504 - val_acc: 0.9846
Epoch 2/5
29404/29404 [==============================] - 14s 475us/step - loss:
0.0755 - acc: 0.9764 - val_loss: 0.0325 - val_acc: 0.9899
Epoch 3/5
29404/29404 [==============================] - 14s 480us/step - loss:
0.0563 - acc: 0.9828 - val_loss: 0.0326 - val_acc: 0.9881
```

```
Epoch 4/5
29404/29404 [==============================] - 14s 480us/step - loss:
0.0472 - acc: 0.9852 - val_loss: 0.0258 - val_acc: 0.9893
Epoch 5/5
29404/29404 [==============================] - 14s 481us/step - loss:
0.0404 - acc: 0.9871 - val_loss: 0.0259 - val_acc: 0.9907
Test score: 0.025926338075212205
Test accuracy: 0.990742645546184
```

Notice that we got 99.8% test accuracy after 5 epochs for the first five digits classifier, and 99.2% for the last five digits after transfer and fine-tuning.

Reinforcement Learning

Reinforcement learning is a goal-oriented learning method based on interaction with its environment. The objective is getting an agent to act in an environment in order to maximize its rewards. Here the agent is an intelligent program, and the environment is the external condition (Figure 6-10).

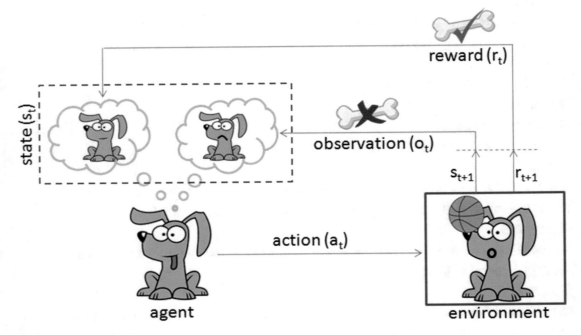

Figure 6-10. *Reinforcement learning is like teaching your dog a trick*

Let's consider an example of a predefined system for teaching a new trick to a dog, where you do not have to tell the dog what to do. However, you can reward the dog if it does the right thing or punish if it does wrong. With every step, it has to remember what made it get reward or punishment; this is commonly known as the credit assignment problem. Similarly, we can train a computer agent such that its objective is to take action to move from state st to state st+1, and find the behavior function to maximize the expected sum of discounted rewards and map the states to actions. According to the paper published by Deepmind Technologies in 2013, the Q-learning rule for updating status is given by: $Q[s,a]_{new} = Q[s,a]_{prev} + \alpha * (r + \gamma*max(s,a) - Q[s,a]_{prev})$, where

α is the learning rate,

r is a reward for the latest action,

γ is the discounted factor, and

max(s,a) is the estimate of new value from best action.

If the optimal value Q[s,a] of the sequence s' at the next time step was known for all possible actions a', then the optimal strategy is to select the action a' maximizing the expected value of $r + \gamma*max(s,a) - Q[s,a]_{prev.}$

Let's consider an example where an agent is trying to come out of a maze (Figure 6-11). It can move one random square or area in any direction and get a reward if it exits. The most common way to formalize a reinforcement problem is to represent it as a Markov decision process. Assume the agent is in state b (maze area) and the target is to reach state f, so within one step the agent can reach from b to f. Let's put a reward of 100 (otherwise 0) for links between nodes that allow agents to reach the target state. Listing 6-17 provides an example code implementation for q-learning.

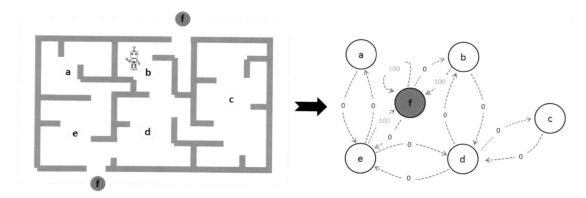

Figure 6-11. *Left: Maze with five states. Right: Markov decision process*

Listing 6-17. Example Code for Q-learning

```
import numpy as np
import matplotlib.pyplot as plt
from matplotlib.collections import LineCollection

# defines the reward/link connection graph
R = np.array([[-1, -1, -1, -1,  0,  -1],
              [-1, -1, -1,  0, -1, 100],
              [-1, -1, -1,  0, -1,  -1],
              [-1,  0,  0, -1,  0,  -1],
              [ 0, -1, -1,  0, -1, 100],
              [-1,  0, -1, -1,  0, 100]]).astype("float32")
Q = np.zeros_like(R)
```

The -1's in the table means there isn't a link between nodes. For example, State 'a' cannot go to State 'b'.

```
# learning parameter
gamma = 0.8

# Initialize random_state
initial_state = randint(0,4)

# This function returns all available actions in the state given as an
argument
def available_actions(state):
    current_state_row = R[state,]
    av_act = np.where(current_state_row >= 0)[1]
    return av_act

# This function chooses at random which action to be performed within the
range
# of all the available actions.
def sample_next_action(available_actions_range):
    next_action = int(np.random.choice(available_act,1))
    return next_action
```

```python
# This function updates the Q matrix according to the path selected and the Q
# learning algorithm
def update(current_state, action, gamma):

    max_index = np.where(Q[action,] == np.max(Q[action,]))[1]

    if max_index.shape[0] > 1:
        max_index = int(np.random.choice(max_index, size = 1))
    else:
        max_index = int(max_index)
    max_value = Q[action, max_index]

    # Q learning formula
    Q[current_state, action] = R[current_state, action] + gamma * max_value

# Get available actions in the current state
available_act = available_actions(initial_state)

# Sample next action to be performed
action = sample_next_action(available_act)

# Train over 100 iterations, re-iterate the process above).
for i in range(100):
    current_state = np.random.randint(0, int(Q.shape[0]))
    available_act = available_actions(current_state)
    action = sample_next_action(available_act)
    update(current_state,action,gamma)

# Normalize the "trained" Q matrix
print ("Trained Q matrix: \n", Q/np.max(Q)*100)

# Testing
current_state = 2
steps = [current_state]

while current_state != 5:
    next_step_index = np.where(Q[current_state,] == np.max(Q[current_
    state,]))[1]
    if next_step_index.shape[0] > 1:
        next_step_index = int(np.random.choice(next_step_index, size = 1))
```

```
    else:
        next_step_index = int(next_step_index)
    steps.append(next_step_index)
    current_state = next_step_index

# Print selected sequence of steps
print ("Best sequence path: ", steps)
#----output----
Best sequence path:  [2, 3, 1, 5]
```

Summary

In this chapter you have learned briefly about various topics of deep learning technique using ANNs, starting from single perceptron, to multilayer perceptron, to more complex forms of deep neural networks such as CNN and RNN. You have learned about the various issues associated with image data and how researchers have tried to mimic the human brain for building models that can solve complex problems related to computer vision and text mining using a convolutional neural network and recurrent neural network, respectively. You also learned how autoencoders can be used to compress/decompress data or remove noise from image data. You learned about the widely popular RBN, which can learn the probabilistic distribution in the input data, enabling us to build better models. You learned about the transfer learning that helps us to use knowledge from one model to another model of similar nature. Finally, we briefly looked at a simple example of reinforcement learning using Q-learning. Congratulations with this! You have reached the end of your six-step expedition of mastering machine learning.

CHAPTER 7

Conclusion

I hope you have enjoyed the six-step, simplified machine learning (ML) expedition. You started your learning journey with step 1, where you learned the core philosophy and key concepts of Python 3 programming language. In step 2 you learned about ML history, high-level categories (supervised/unsupervised/reinforcement learning), and three important frameworks for building ML systems (SEMMA, CRISP-DM, KDD data mining process), primary data analysis packages (NumPy, Pandas, Matplotlib) and their key concepts, and comparison of different core ML libraries. In step 3 you learned different data types, key data quality issues and how to handle them, exploratory analysis, and core methods of supervised/unsupervised learning and their implementation with an example. In step 4 you learned the various techniques for model diagnosis, bagging for overfitting, boosting for underfitting, ensemble techniques; and hyperparameter tuning (grid/random search) for building efficient models. In step 5 you got an overview of the text mining process: data assembled, data preprocessing, data exploration or visualization, and various models that can be built. You also learned how to build collaborative/content-based recommender systems to personalize user experience. In step 6 you learned about artificial neural networks through perceptron, convolutional neural networks (CNNs) for image analytics, recurrent neural networks (RNNs) for text analytics, and a simple toy example for learning the reinforcement learning concept. These are the advanced topics that have seen great development in the last few years.

Overall, you have learned a broad range of commonly used ML topics; each of them come with a number of parameters to control and tune model performance. To keep it simple throughout the book, I have either used the default parameters or you were introduced only to the key parameters (in some places). The default options for parameters have been carefully chosen by the creators of the packages to give decent results to get you started. So, to start with you can go with the default parameters. However, I recommend you to explore the other parameters and play

© Manohar Swamynathan 2019
M. Swamynathan, *Mastering Machine Learning with Python in Six Steps*,
https://doi.org/10.1007/978-1-4842-4947-5_7

with them using a manual/grid/random search to ensure a robust model. Table 7-1 is a summary of various possible problem types, example use cases, and the potential ML algorithms that you can use. Note that this is a sample list only, not an exhaustive list.

Table 7-1. *Problem Type vs. Potential ML Algorithms*

Problem Type	Example Use Case(s)	Potential ML Algorithms
Predicting a continuous number	What will be store daily/weekly sales?	Linear regression or polynomial regression
Predicting a count type of continuous number	How many staff are required for a shift? How many car park spaces are required for a new store?	Generalized linear model with a Poisson distribution
Predict the probability of an event (True/False)	What is the probability of a transaction being a fraud?	Binary classification models (logistic regression, decision tree models, boosting models, kNN, etc.)
Predict the probability of event out of many possible events (multiclass)	What is the probability of a transaction being high risk/medium risk/low risk?	Multiclass classification models (logistic regression, decision tree models, boosting models, kNN, etc.)
Group the contents based on similarity	Group similar customers? Group similar categories?	k-means clustering, hierarchical clustering
Dimension reduction	What are the important dimensions that hold a maximum percentage of information?	Principal component analysis (PCA), singular value decomposition (SVD)
Topic Modeling	Group documents based on topics or thematic structure?	Latent Dirichlet allocation, nonnegative matrix factorization
Opinion Mining	Predict the sentiment associated with the text?	Natural Language Toolkit (NLTK)
Recommend systems	What products/items to be marketed to a user?	Content-based filtering, collaborative filtering

(continued)

Table 7-1. (*continued*)

Problem Type	Example Use Case(s)	Potential ML Algorithms
Text classification	Predict the probability of a document being part of a known class?	Recurrent neural network (RNN), binary or multiclass classification models
Image classification	Predict the probability of the image being part of a known class?	Convolutional neural network (CNN), binary or multiclass classification models

Tips

Building an efficient model can be a challenging task for a starter. Now that you have learned what algorithm to use, I would like to give my 2 cents list of things to remember while you get started on the model building activity.

Start with Questions/Hypothesis, Then Move to Data!

Don't jump into understanding the data before formulating the objective to be achieved using he data. It is a good practice to start with a list of questions and work closely with domain experts to understand the core issue and frame the problem statement. This will help you in choosing the right ML algorithm (supervised vs. unsupervised) and then move on to understanding different data sources (Figure 7-1).

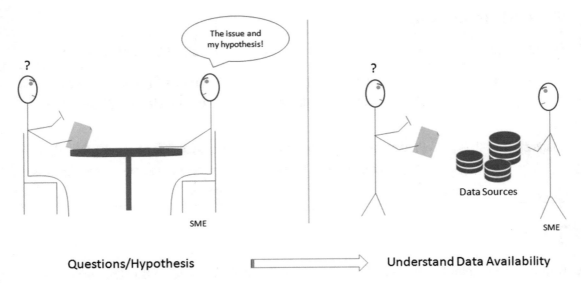

Figure 7-1. *Questions/hypothesis to data*

Don't Reinvent the Wheel from Scratch

The ML open source community is very active; there are plenty of efficient tools available, and many more are being developed/released often. So do not try to reinvent the wheel in terms of solutions/algorithms/tools unless required (Figure 7-2). Try to understand what solutions exist in the market before venturing into building something from scratch.

Figure 7-2. *Don't reinvent the wheel*

Start with Simple Models

Always start with simple models (such as regressions), as these can be explained easily in layman's terms to any non-techie people (Figure 7-3). This will help you and the subject matter experts to understand the variable relationships and gain confidence in the model. Further, it will significantly help you to create the right features. Move to complex models only if you see a noteworthy increase in the model performance.

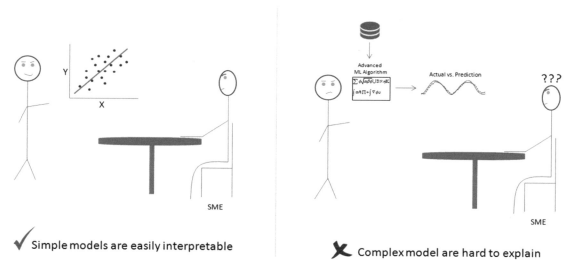

Figure 7-3. Start with a simple model

Focus on Feature Engineering

Relevant features lead to efficient models, not more features! Note that including a large number of features might lead to an overfitting problem. Including relevant features in the model is the key to building an efficient model. Remember that the feature engineering part is talked about as an art form and is the key differentiator in competitive ML. The right ingredients mixed to the right quantity are the secret for tasty food; similarly, passing the relevant/right features to the ML algorithm is the secret to an efficient model (Figure 7-4).

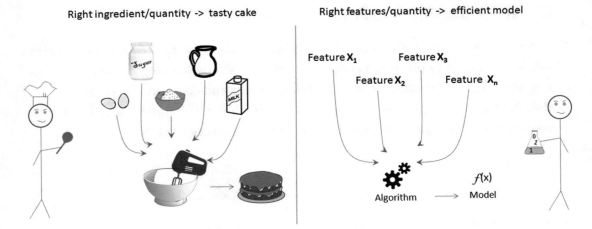

Figure 7-4. *Feature engineering is an art*

Beware of Common ML Imposters

Carefully handle some of the common ML imposters such as data quality issues (such as missing data, outliers, categorical data, scaling), the imbalanced dataset for classification, overfitting, and underfitting. Use the appropriate techniques discussed in Chapter 3 for handling data quality issues and techniques discussed in Chapter 4, such as ensemble techniques and hyperparameter tuning, to improve the model performance.

Happy Machine Learning

I hope this expedition into machine learning in simplified six steps has been worthwhile, and I hope this helps you to start a new journey of applying them on real-world problems. I wish you all the very best and success for your further quest.

Index

A

add() method, 41
Adjusted R-squared, 185
Arithmetic operators, 16
Artificial general intelligence
        (AGI), 71, 383
Artificial intelligence evolution, 65
   AGI, 71
   ANI, 71
   ASI, 71
   Bayes' theorem, 73
   data analytics, 82
      descriptive, 78
      diagnostic, 78
      ERP, 76
      predictive, 79
      prescriptive, 79
      types, 76
   data mining, 75, 82
   data science, 80–82
   frequentist, 73
   regression method, 74
   statistics, 72, 82
Artificial Narrow Intelligence (ANI), 71
Artificial Neural Network (ANN)
   classification model, 387
   handwritten digit, 386
   visual pathway, 385, 386
Artificial Super Intelligence (ASI), 71
Assignment operators, 19

Autoencoder
   data denoising, 407
   denoise, 413, 414
   dimension reduction, 408–413
   elements, 407, 408
Autoregressive integrated moving
        average (ARIMA)
   AM, 234
   autocorrelation test, 238, 239
   check, 236, 237
   decompose time
      series, 235, 236
   first order differencing, 241, 242
   key parameters, 234
   MA, 234
   model and evaluate, 239, 240
   optimal parameter, 235
   predict function, 243
Autoregressive model (AM), 234

B

Bag of words (BoW), 347
Bayesian optimization
   additional resources, 323
   noise reduction, 319
   random search, 317, 318
Bayes' theorem, 73
Bias, 274
Bitwise operators, 20

© Manohar Swamynathan 2019
M. Swamynathan, *Mastering Machine Learning with Python in Six Steps*,
https://doi.org/10.1007/978-1-4842-4947-5

Printed in the United States
By Bookmasters